ADVANCED ELECTRICAL INSTALLATION WORK

Second Edition

ADVANCED

ELECTRICAL INSTALLATION WORK

Second Edition

TREVOR LINSLEY
Senior Lecturer, Blackpool and The Fylde College

Edward Arnold
A member of the Hodder Headline Group
LONDON MELBOURNE AUCKLAND

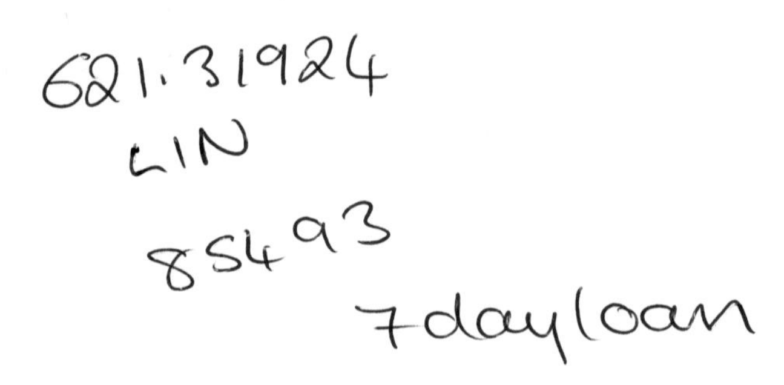

To Joyce, Samantha and Victoria

Edward Arnold is a division of Hodder Headline PLC
338 Euston Road, London NW1 3BH

First published in the United Kingdom 1989
Second edition 1992

British Library Cataloguing in Publication Data
Linsley, Trevor
Advanced Electrical Installation Work.-
2Rev.ed
I. Title
621.31924

ISBN 0 340 56833 X

4 6 7 5 3
94 96 98 97 95

Typeset in Times Roman by
Wearset, Boldon, Tyne and Wear
Printed and bound in the United Kingdom by
The Bath Press, Avon

PREFACE

This book has been written as a complete textbook for the City & Guilds 236 Electrical Installation Competences Part 2. The book meets the combined requirements of the Installation Practice and Technology Competences of the 236 Syllabus and therefore students need purchase only this textbook for all subjects in the Part 2 course.

This second volume in the series seeks to deepen the student's understanding of the craft processes and the associated technology with which the electrical contracting industry now operates. The book will also provide a sound basic knowledge to other professionals in the construction and electrical service industries and to those students taking BTEC and SCOTVEC Electrical and Utilisation Units.

Modern Regulations place a greater responsibility upon the installing electrician for safety and the design of an installation. The latest Regulations governing electrical installations are the 16th Edition of the IEE Wiring Regulations. The second edition of this book has been revised and updated to incorporate the 16th Edition of the IEE Wiring Regulations.

The Part 2 examination in Electrical Installation Competences comprises one multiple choice and one written examination paper. Multiple choice and written questions are included at the end of each chapter as an aid to self study and to encourage a thorough knowledge of the subjects.

I would like to acknowledge the assistance given by the following manufactures and organisations in the preparation of this book.

Crabtree Electrical Industries Limited
Thorn Lighting Limited
The CEGB
Avo Instruments
The Institution of Electrical Engineers
The British Standards Institution
The City & Guilds of London Institute

I would also like to thank my colleagues and students at Blackpool & the Fylde College for giving me the motivation to write this book.

Finally I would like to thank Joyce, Samantha and Victoria for their support and encouragement.

Trevor Linsley,
Poulton-le-Fylde 1992.

CONTENTS

CHAPTER 1

Regulations

Electricity generation as we know it today began when Michael Faraday conducted the famous ring experiment in 1831. This experiment, together with many other experiments of the time, made it possible for Lord Kelvin and Sebastian de Ferranti to patent in 1882 the designs for an electrical machine called the Ferranti–Thompson dynamo, which made possible the generation of electricity on a commercial scale.

In 1887 the London Electric Supply Corporation was formed with Ferranti as the Chief Engineer. This was one of many privately owned electricity generating stations supplying the electrical needs of the UK. As the demand for electricity grew, more privately owned generating stations were built until eventually the Government realised that electricity was a national asset which would benefit from nationalisation.

In 1926 the Electricity Supply Act placed the responsibility for generation in the hands of a Central Electricity Board. In England and Wales the Central Electricity Generating Board (CEGB) had the responsibility of the generation and transmission of electricity on the Supergrid. In Scotland, generation was the joint responsibility of the North of Scotland Hydro-Electricity Board and the South of Scotland Electricity Board. In Northern Ireland electricity generation was the responsibility of the Northern Ireland Electricity Service.

In 1988 Mr. Cecil Parkinson, the Secretary of State for Energy in the Conservative Government proposed the de-nationalisation of the electricity supply industry, returning the responsibility for generation, transmission and distribution to the private sector. It is anticipated that this action, together with new legislation over the security of supplies, will lead to a guaranteed quality of the provision, with increased competition leading eventually to cheaper electricity.

During the period of development of the electricity services, particularly in the early days, poor design and installation led to many buildings being damaged by fire and the electrocution of human beings and livestock. It was the insurance companies who originally drew up a set of rules and guidelines of good practice in the interest of reducing the number of claims made upon them. The first rules were made by the American Board of Fire Underwriters and were quickly followed by the Phoenix Rules of 1882. In the same year the first edition of the Rules and Regulations for the Prevention of Fire Risk arising from Electrical Lighting were issued by the Institute of Electrical Engineers.

The current edition of these Regulations is called the Regulations for Electrical Installations and from January 1993 we will be using the 16th Edition. All the rules have been revised, updated and amended at regular intervals to take account of modern developments and the 16th Edition brings the U.K. Regulations into harmony with Europe. The electrical industry is now controlled by at least five sets of rules, regulations and standards. These are:

- the Electricity Supply Regulations, 1988,
- the Electricity at Work Regulations 1989,
- the IEE Regulations for Electrical Installations,
- British Standards,
- the Health and Safety at Work Act, 1974.

The Electricity Supply Regulations are issued by the Department of Trade and Industry. They are statutory regulations which are enforceable by the laws of the land. They are designed to ensure a proper and safe supply of electrical energy up to the consumer's terminals. The Regional Electricity Companies must declare the supply voltage and maintain it within 6% of the value indicated. The frequency is maintained at an average value of 50 Hz over 24 hours so that electric clocks remain accurate.

Regulation 29 gives the area boards the power to refuse to connect a supply to an installation which in their opinion is not constructed, installed and protected to an appropriately high standard. This regulation would only be enforced if the installation did not meet the requirements of the IEE Regulations for Electrical Installations.

The Electricity at Work Regulations 1989 came into force on the 1st April 1990. The Regulations are made under the Health & Safety at Work Act 1974, and enforced by the Health & Safety Executive. The purpose of the Regulations is to reduce the risk of death or personal injury from electricity in the workplace.

Electrical installations which have been installed in accordance with the IEE Regulations are likely to achieve compliance with the Electricity at Work Regulations.

Regulation 4 of the 1989 Regulations is an important one for electrical contractors. It states that all systems, work activities and protective equipment shall be constructed, maintained and carried out in a manner which will not give rise to danger. 'Live working' is probably the most dangerous part of an electrician's working activities and must now be avoided at all times. The Regulations tell us that equipment must be isolated and securely locked off and electrical conductors made dead before work starts. Work must only be carried out by a 'competent' person. Regulation 13 tells us that *'formalised house rules'* or *'permits to work'* may form part of the written procedures considered essential to ensure safe working.

This means that electricians must be trained in, for example, isolation procedures and that records of training and documented procedures must be kept by the employer so that he can if necessary, legally 'prove' compliance with the 1989 Regulations.

Electrical contractors must become more 'legally aware' following the conviction of an electrician for manslaughter at Maidstone Crown Court in 1989. The court accepted that an electrician caused the death of another man as a result of his shoddy work in wiring up a central heating system. He received a nine month suspended prison sentence. This case has set an important legal precedent and in future any tradesman or professional who causes death through negligence or poor workmanship risks prosecution for manslaughter and possible imprisonment.

The Institution of Electrical Engineers Regulations for Electrical Installations is the electrician's 'bible'. The Regulations provide an authoritative set of requirements for the design and installation of electrical systems in buildings. The use of the word 'Regulations' in the IEE document does not indicate that it contains legal regulations which are enforceable by law, but failure to comply with the IEE Regulations will contravene one of the other Statutory Regulations, which are enforceable by law. The IEE Regulations are held in such high regard by the statutory bodies that they have refrained from publishing any competitive regulations. Therefore, the IEE Regulations for Electrical Installations are for all practical purposes the national standard for electrical installation practice within the UK.

British Standards indicate that goods manufactured to the exacting specifications laid down by the British Standards Institution are suitable for the purpose for which they were made. There seems to be a British Standard for practically everything made today, and compliance with the relevant British Standard is in most cases voluntary. However, when specifying or installing equipment, the electrical designer or contractor needs to be sure that the materials are suitable for their purpose and offer a degree of safety, and should only use equipment which carries the appropriate British Standards number.

The British Standards Institute have created two important marks of safety; the BSI kite mark and the BSI safety mark, which are shown in Figure 1.1.

The BSI kite mark is an assurance that the product carrying the label has been produced under a system of supervision, control and testing and can only be used by manufacturers who have been granted a licence under the scheme. It does not necessarily cover safety unless the appropriate British Standard specifies a safety requirement.

The BSI safety mark is a guarantee of the product's electrical, mechanical and thermal safety. It does not guarantee the product's performance.

The IEE Regulations make specific reference to many British Standard Specifications and Brit-

BSI kite mark

BSI safety mark

Figure 1.1 BSI safety marks

ish Standard Codes of Practice in the 16th Edition of the Regulations.

The Electricity Supply Regulations forbid electricity supply authorities to connect electric lines and apparatus to the supply system unless their insulation is capable of withstanding the tests prescribed by the appropriate British Standards.

The Health and Safety at Work Act provides a legal framework for stimulating and encouraging high standards of health and safety for everyone at work and the public at large from risks arising from work activities. The Act was a result of recommendations made by a Royal Commission in 1970 which looked at the health and safety of employees at work, and concluded that the main cause of accidents was apathy on the part of the employer and employee. The new Act places the responsibility for safety at work on both the employer and the employee.

Section 2 of the Act details what the employer must do to comply with the legal requirements of the Act. He must, for example, provide adequate maintenance of plant and equipment and provide a safe working environment with appropriate protective equipment so that the worker can carry out his duties safely.

Section 7 of the Act details what the employee must do to comply with the Act. These include taking reasonable care to avoid injury to himself or others as a result of his work activities, and co-operating with his employer in helping him to comply with the Act.

The Act is enforced by the Health and Safety Executive who also provide an advisory service to both employer and employee organisations. The Inspector has the authority to enter any premises, collect evidence and issue improvement or prohibition notices. The Act provides for criminal proceedings to be instituted against those who do not satisfy the requirements of the Regulations.

It is clearly in the interests of the electrical contractor to be aware of, and to comply with, any regulations which are relevant to the particular Installation. IEE Regulation 130–01 states that good workmanship and the use of proper materials are essential for compliance with the Regulations.

In order to try to ensure that all electrical installation work is carried out to a minimum standard, the National Inspection Council for Electrical Installation Contracting (NICEIC) was established in 1956. The NICEIC is supported by all sections of the electrical industry and its aims are to provide consumers with protection against faulty, unsafe or otherwise defective electrical installations. The council maintains an approved roll of members who regularly have their premises, equipment and installations inspected by the council's engineers. Through this inspectorate the council is able to ensure a minimum standard of workmanship among its members. The electricians employed by an NICEIC approved contractor are also, by association with their employer, accepted as being competent to carry out electrical installation work to an approved standard.

Exercises

1. Equipment which displays the British Standard Kite mark:
   (a) is guaranteed to perform efficiently, (b) has been produced under a system of supervision and control by a manufacturer holding a licence, (c) will reduce the risk of an electric shock, under fault conditions, to anyone using the product, (d) carries a guarantee of the product's electrical, mechanical and thermal safety.
2. Equipment which carries the British Standard safety mark:
   (a) is guaranteed to perform efficiently, (b) has been produced under a system of supervision and control by a manufacturer holding a licence, (c) will reduce the risk of an electric shock, under fault conditions, to anyone using the product, (d) carries a guarantee of the products electrical, mechanical and thermal safety.
3. Make a list of five things which an electrician must do at work in order to comply with the Health and Safety at Work Act.
4. What are the IEE Regulations and how do they influence the work of an electrician?
5. Who are the NICEIC and how do they seek to maintain high standards in the electrical contracting industry?

CHAPTER 2

Generation, transmission and distribution of electricity

Introduction

Following the de-nationalisation of the electricity supply industry, the generation of electrical energy in Great Britain has become the responsibility of three generating companies formed from the previously nationalised Central Electricity Generating Board. These are known as National Power, PowerGen and Nuclear Electric. National Power was allocated 40 of the power stations previously operated by the CEGB and provides about 48% of the electricity supply in England and Wales. They have a generating capacity of approximately 30 000 MW, 80% of the power stations are coal fired and 20% oil fired.

PowerGen was allocated 21 of the CEGB power stations and provides about 30% of the electricity in England and Wales. They have a generating capacity of approximately 19 000 MW, 80% of the power stations are coal fired and 20% oil fired.

Nuclear Power remains a state owned company which will operate all of the nuclear power stations. They will provide between 15% and 20% of the daily base load power from 14 power stations.

The twelve Electricity Board Areas have become the Regional Electricity Companies who own the National Grid Company and are responsible for the distribution of electricity to individual homes, offices and industry.

The North of Scotland Hydro-Electricity Board and the South of Scotland Electricity Board have been renamed Hydro-Electric and Scottish Power.

These two Scottish electricity companies generate, transmit, distribute and sell electricity from the power station to the individual consumer.

In England and Wales, the three generating companies generate the power, the National Grid Company transmits it to the twelve Regional Electricity Companies who distribute and sell to the individual consumers.

Figure 2.1 shows the twelve Electricity Board Areas which are now called the Regional Electricity Companies.

Generation of electricity in most modern power stations is at 25 kV and this voltage is then transformed to 400 kV for transmission. Virtually all the generators of electricity throughout the world are three phase synchronous generators. The generator consists of a prime mover and a magnetic field excitor. The magnetic field is produced electrically by passing a direct current through a winding on an iron core, which rotates inside three phase windings on the stator of the machine. The magnetic field is rotated by means of a prime mover which may be a steam turbine, water turbine or gas turbine. Primary sources for electricity generation are discussed in Chapter 1 of *Basic Electrical Installation Work*.

The generators in modern power stations are rated between 500 MW and 1000 MW. A 2000 MW station might contain four 500 MW sets, three 660 MW sets and a 20 MW gas turbine generator or two 1000 MW sets. Having a number of generator sets in a single power station provides the flexibility required for seasonal variations in the load and for maintenance of equipment. When generators are connected to a single system they must rotate at exactly the same speed, hence the term synchronous generator.

Very high voltages are used for transmission systems because, as a general principle, the higher the voltage the cheaper is the supply. Since

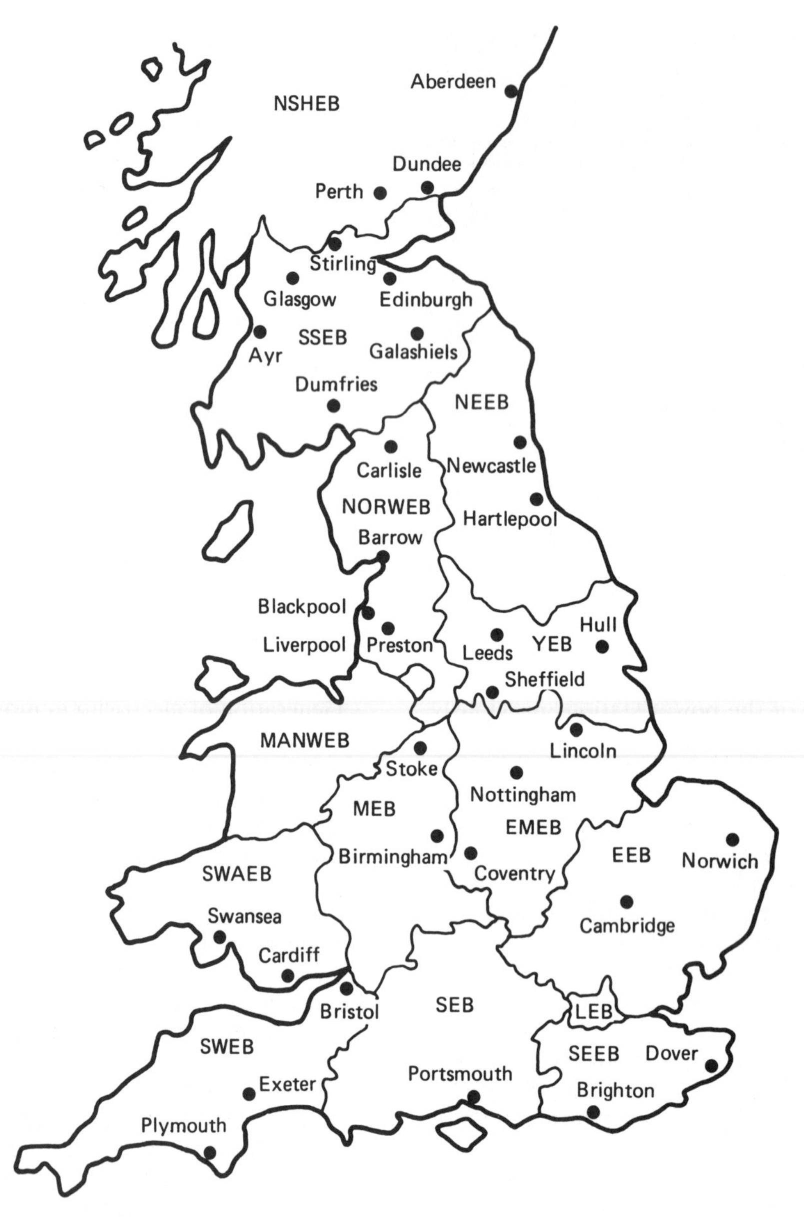

Figure 2.1 Electricity Board areas in the United Kingdom now called the Regional Electricity Companies.

$P = VI \cos \theta$, it follows that an increase in voltage will reduce the current for a given amount of power. A lower current will result in reduced cable and switchgear size and the line power losses, given by the equation $P = I^2 R$, will also be reduced.

The 132 kV grid and 400 kV Super Grid transmission lines are, for the most part, steel cored aluminium conductors suspended on steel lattice towers, since this is about 16 times cheaper than the equivalent underground cable. Figure 2.2 shows a suspension tower on the Sizewell–Sundon 400 kV transmission line. The conductors are attached to porcelain insulator strings which are fixed to the cross-members of the tower as shown in Figure 2.3. Three conductors com-

Figure 2.2 Suspension tower on Sizewell–Sundon transmission line (by kind permission of the CEGB)

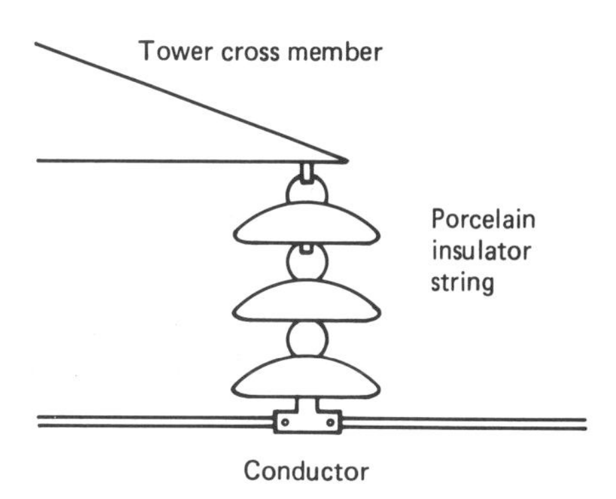

Figure 2.3 Steel latice tower cable supports

prise a single circuit of a three phase system so that towers with six arms carry two separate circuits.

Primary distribution to consumers is from 11 kV sub-stations, which for the most part are fed from 33 kV sub-stations, but direct transformation between 132 kV and 11 kV is becoming common policy in city areas where over 100 MW can be economically distributed at 11 kV from one site. Figure 2.4 shows a block diagram indicating the voltages at the various stages of the transmission and distribution system.

Distribution systems at 11 kV may be ring or radial systems but a ring system offers a greater security of supply. The maintenance of a secure supply is an important consideration for any electrical engineer or supply authority because electricity plays a vital part in an industrial society, and a loss of supply may cause inconvenience, financial loss or danger to the consumer or the public.

The principle employed with a ring system is that any consumer's sub-station is fed from two directions, and by carefully grading the overload and cable protection equipment a fault can be disconnected without loss of supply to other consumers.

High voltage distribution to primary sub-stations is used by the electricity boards to supply small industrial, commercial and domestic consumers. This distribution method is also suitable for large industrial consumers where 11 kV sub-stations, as shown in Figure 2.5, may be strategically placed at load centres around the factory site. Regulation 9 of the Electricity Supply Regulations and Regulation 31 of the Factories Act require that these sub-stations be protected by 2.44 m high fences or enclosed in some other way so that no unauthorised person may gain access to the potentially dangerous equipment required for 11 kV distribution. In towns and cities the sub-station equipment is usually enclosed in a brick building as shown in Figure 2.6.

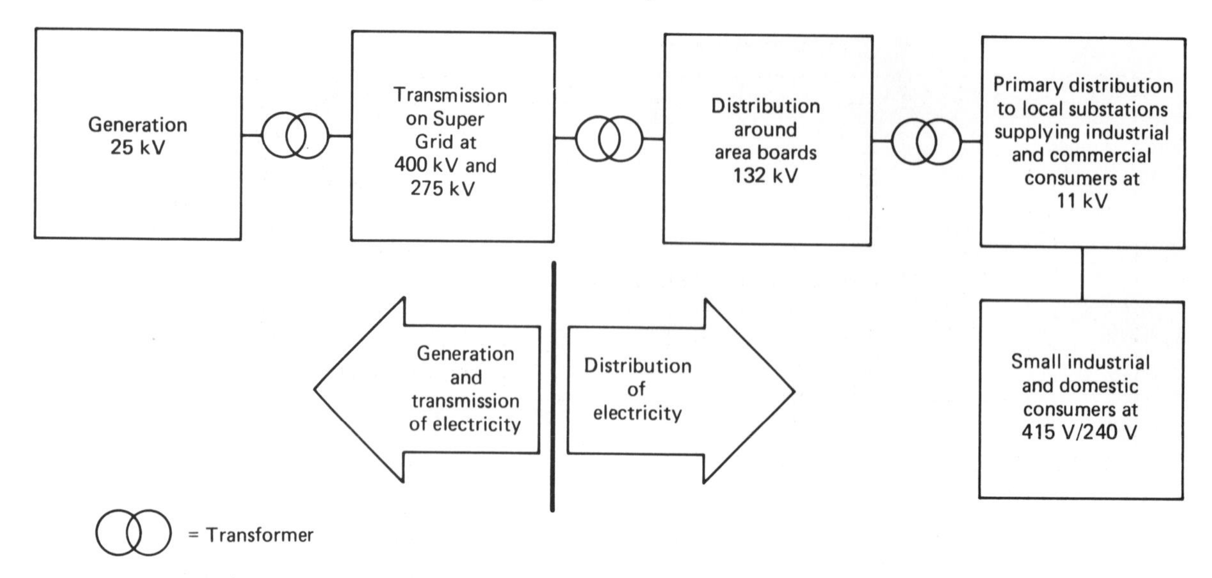

Figure 2.4 Generation, transmission and distribution of electrical energy

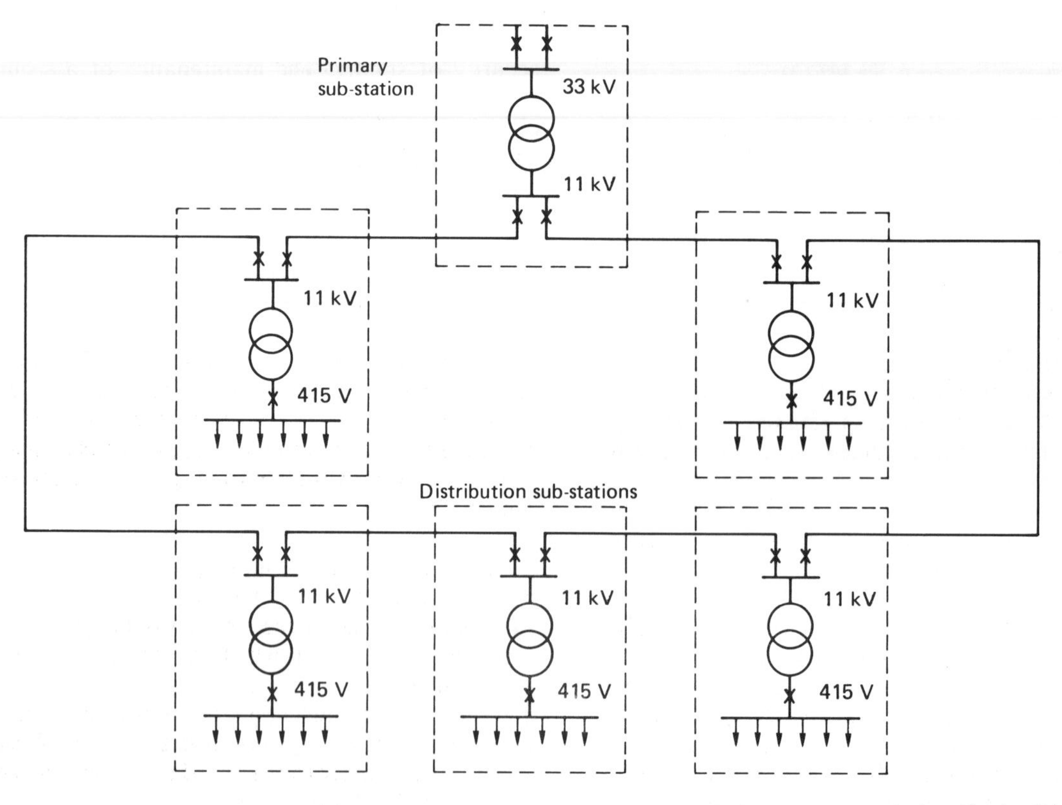

Figure 2.5 High voltage ring main distribution

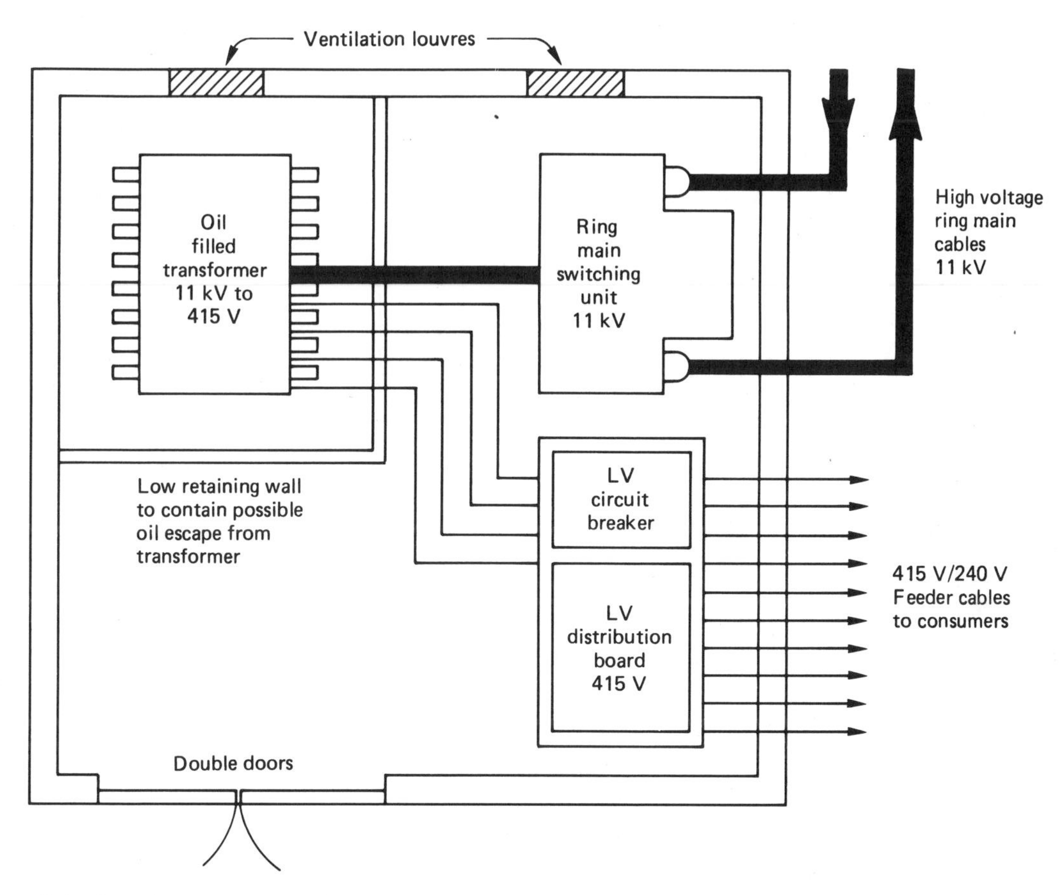

Figure 2.6 Typical sub-station layout

The final connections to plant, distribution boards, commercial or domestic loads are usually by simple underground radial feeders at 415 V/240 V. These outgoing circuits are usually protected by circuit breakers in a distribution board.

The 415 V/240 V is derived from the 11 kV/415 V sub-station transformer by connecting the secondary winding in star as shown in Figure 2.7. The star point is earthed to an earth electrode sunk into the ground below the sub-station, and from this point is taken the fourth conductor, the neutral. Loads connected between phases are fed at 415 V, and those fed between one phase and neutral at 240 V. A three phase 415 V supply is used for supplying small industrial and commercial loads such as garages, schools and blocks of flats. A single phase 240 V supply is usually provided for individual domestic consumers.

Example

Use a suitable diagram to show how a 415 V three phase, four wire supply may be obtained from an 11 kV delta-connected transformer. Assuming that the three phase four wire supply feeds a small factory, show how the following loads must be connected:

a) a three phase 415 V motor,
b) a single phase 415 V welder,
c) a lighting load made up of discharge lamps arranged in a way which reduce the stroboscopic effect.
d) State why 'balancing' of loads is desirable.
e) State the advantages of using a three phase four wire supply to industrial premises instead of a single phase supply.

Figure 2.7 shows the connections of the 11 kV to 415 V supply and the method of connecting a

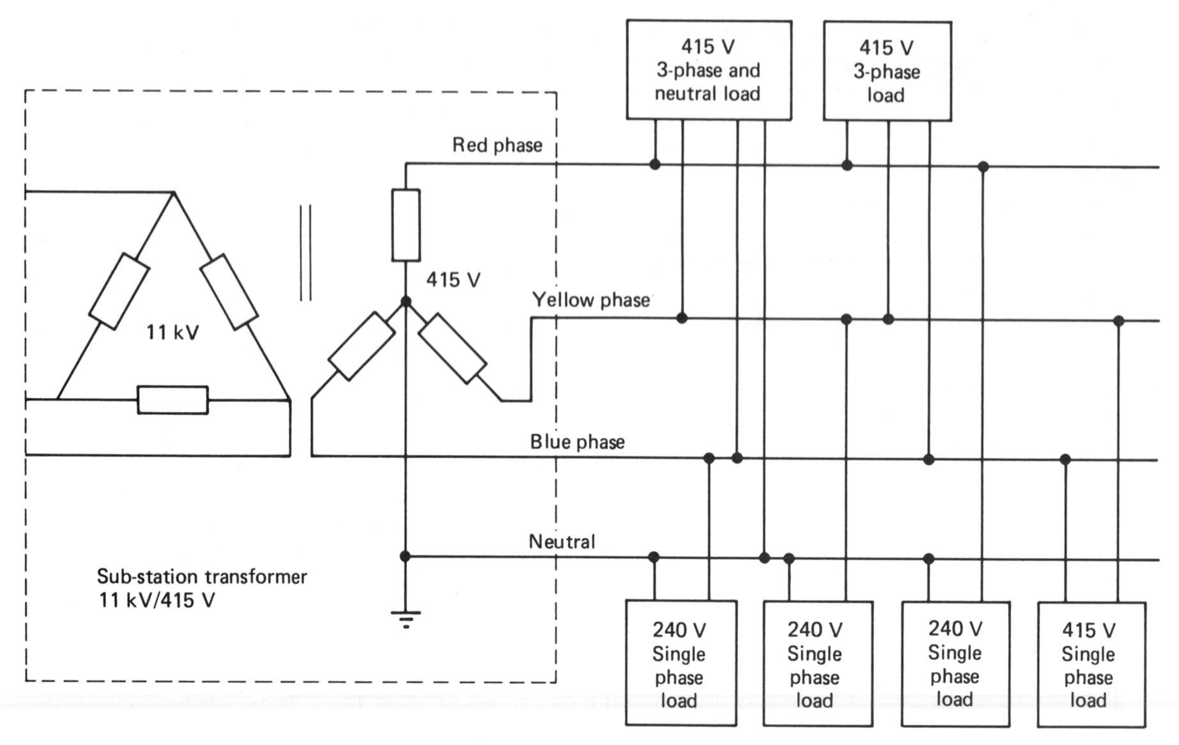

Figure 2.7 Three phase four wire distribution

415 V three phase load such as a motor and a 415 V single phase load such as a welder.

The stroboscopic effect may be reduced by equally dividing the lighting load across the three phases of the supply. For example, if the lighting load were made up of 18 luminaires, then six luminaires should be connected to the red phase and neutral, six to the blue phase and neutral and six to the yellow phase and neutral. (Stroboscopic effect and its elimination are discussed in some detail in Chapter 8 of *Basic Electrical Installation Work*).

A three phase load such as a motor has equally balanced phases since the resistance of each phase winding will be the same. Therefore the current taken by each phase will be equal. When connecting single phase loads to a three phase supply, care should be taken to distribute the single phase loads equally across the three phases so that each phase carries approximately the same current. Equally distributing the single phase loads across the three phase supply is known as 'balancing' the load. A lighting load of 18 luminaires would be 'balanced' if six luminaires were connected to each of the three phases.

A three phase four wire supply gives a consumer the choice of a 415 V three phase supply and a 240 V single phase supply. Many industrial loads such as motors require a three phase 415 V supply, whilst the lighting load in a factory, as in a house, will be 240 V. Industrial loads usually demand more power than a domestic load, and more power can be supplied by a 415 V three phase supply than is possible with a 240 V single phase supply for a given size of cable since Power = VI cos θ (Watts).

Low voltage supply systems

A domestic, commercial or small industrial consumer's installation is usually protected at the incoming service cable position with a 100 A HBC fuse. Other equipment at this position are the energy meter and the consumer's distribution unit, providing the protection for the final circuits and the earthing arrangements for the installation.

An efficient and effective earthing system is essential to allow protective devices to operate. The limiting values of earth fault loop impedance are given in Tables 41, 604 and 605 of the IEE Regulations, and Section 542 gives details of the earthing arrangements to be incorporated in the supply system to meet the requirements of the Regulations. Five systems are described in the definitions but only the TN-S, TN-C-S and TT systems are suitable for public supplies.

A system consists of an electrical installation connected to a supply. Systems are classified by a capital letter designation.

The supply earthing

Arrangements are indicated by the first letter where T means one or more points of the supply are directly connected to earth and I means the supply is not earthed or one point is earthed through a fault limiting impedance.

The installation earthing

Arrangements are indicated by the second letter where T means the exposed conductive parts are connected directly to earth and N means the exposed conductive parts are connected directly to the earthed point of the source of the electrical supply.

The earthed supply conductor

Arrangement is indicated by the third letter where S means a separate neutral and protective conductor and C means that the neutral and protective conductors are combined in a single conductor.

TN-S system

This is one of the most common types of supply system to be found in the UK where the Electricity Board's supply is provided by underground cables. The neutral and protective conductor are separate throughout the system. The protective conductor (PE) is the metal sheath and armour of the underground cable and this is connected to

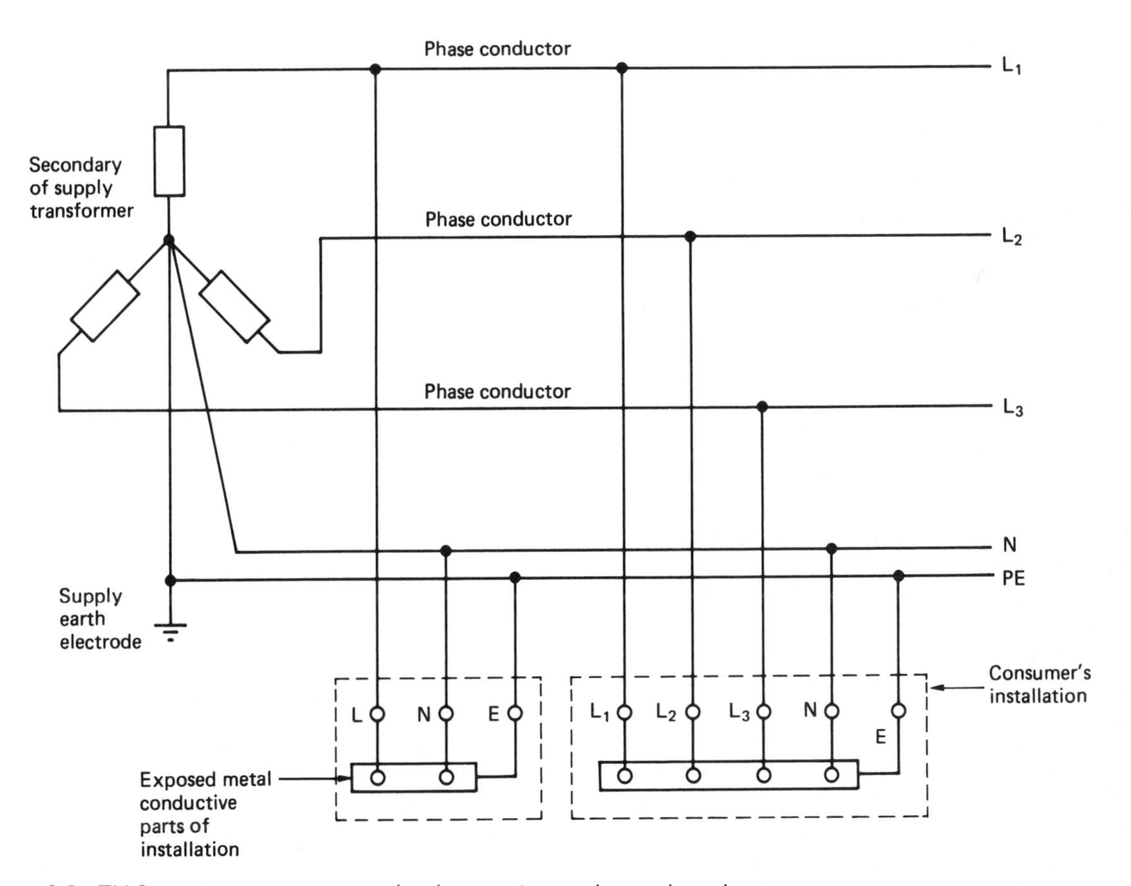

Figure 2.8 TN-S system – separate neutral and protective conductor throughout

the consumer's main earthing terminal. All exposed conductive parts of the installation, gas pipes, water pipes and any lightning protective system are connected to the protective conductor via the main earthing terminal of the installation. The arrangement is shown in Figure 2.8.

TN-C-S system

This type of underground supply is becoming increasingly popular to supply new installations in the UK. It is more commonly referred to as protective multiple earthing (PME). The supply cable uses a combined protective and neutral conductor (PEN conductor). At the supply intake point a consumer's main earthing terminal is formed by connecting the earthing terminal to the neutral conductor. All exposed conductive parts of the installation, gas pipes, water pipes and any lightning protective system are then connected to the main earthing terminals. Thus phase to earth faults are effectively converted into phase to neutral faults. The arrangement is shown in Figure 2.9.

TT system

This is the type of supply more often found when the installation is fed from overhead cables. The supply authorities do not provide an earth terminal and the installation's circuit protective conductors must be connected to earth via an earth electrode provided by the consumer. An effective earth connection is sometimes difficult to obtain and in most cases a residual current device is provided when this type of supply is used. The arrangement is shown in Figure 2.10.

The layout of a typical domestic service position for these three supply systems is shown in Figure 2.11. The TN-C and IT systems of supply do not comply with the Supply Regulations and therefore cannot be used for public supplies. Their use is restricted to private generating plants.

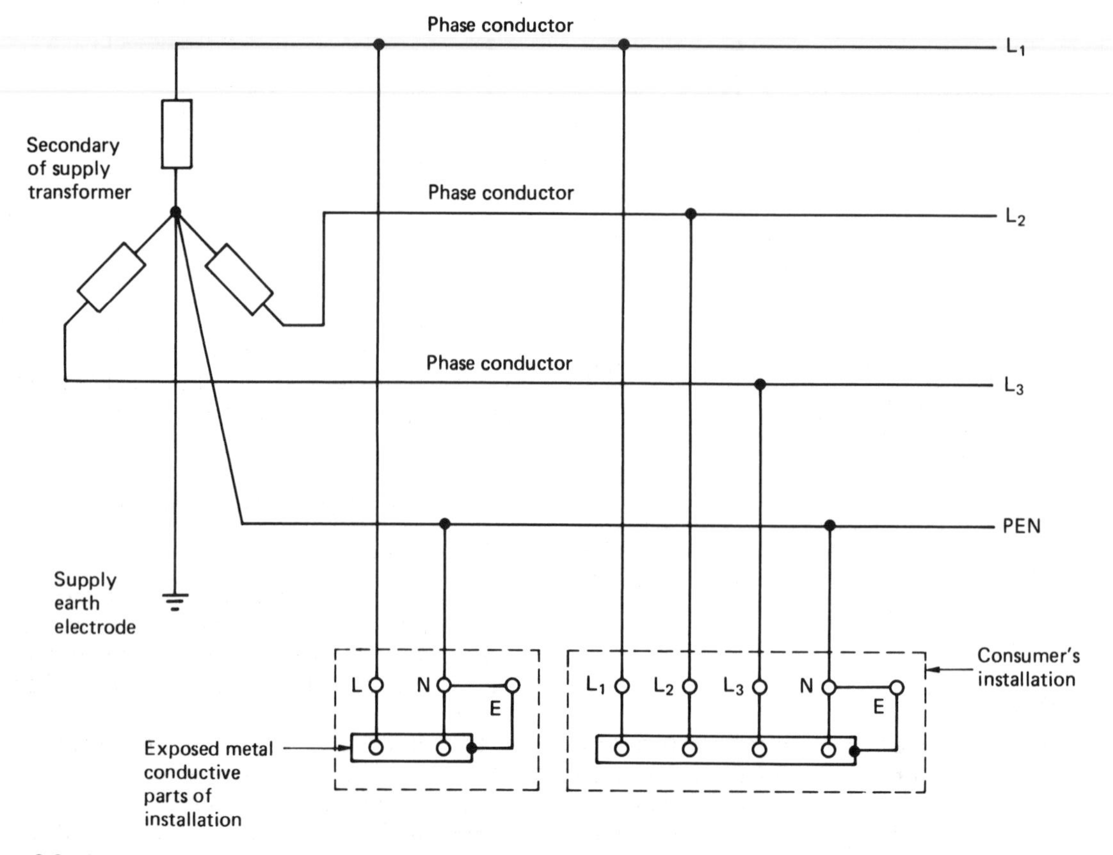

Figure 2.9 TN-C-S system – neutral and protective functions combined in a single (PEN) conductor

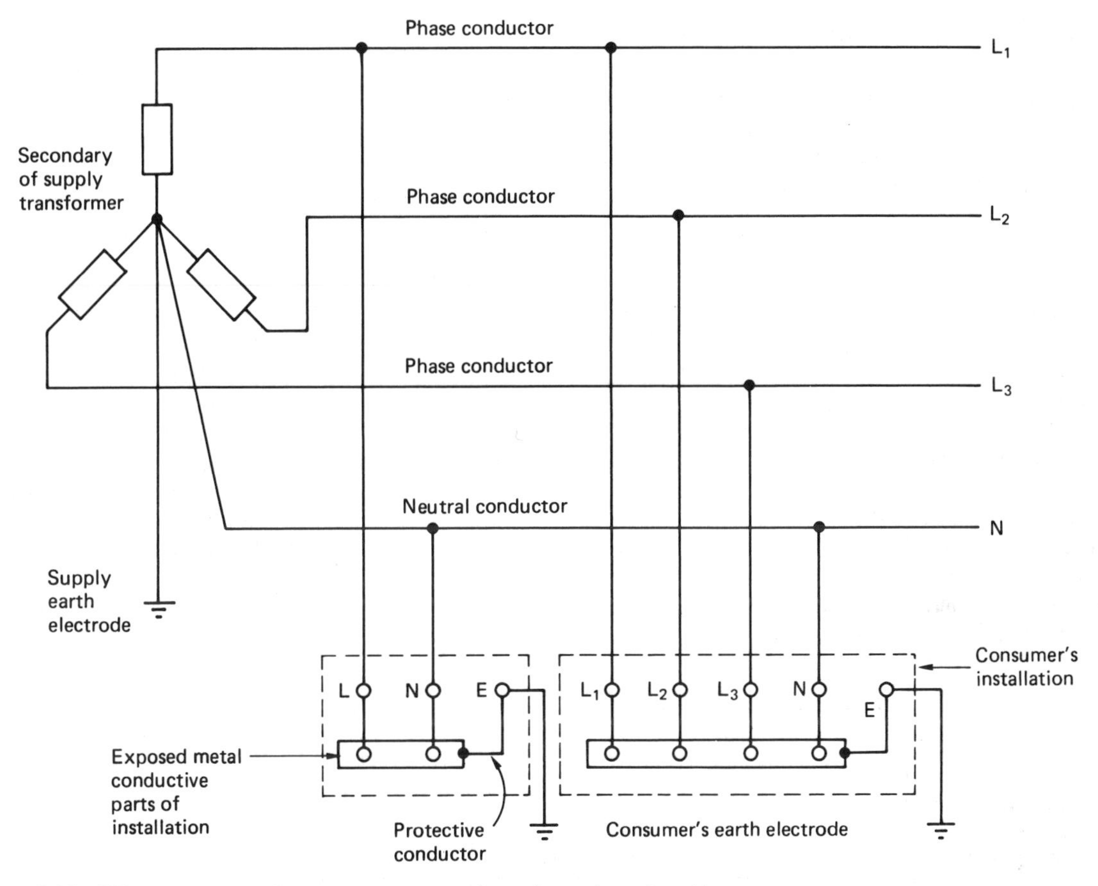

Figure 2.10 TT systems – earthing arrangements independent of supply cable

TN-C system

The supply cable and the consumer's installation use a combined protective and neutral conductor (PEN conductor). All exposed conductive parts of an installation are connected to the PEN conductor as shown in Figure 2.12. The applications of this supply system are limited to privately owned generating plants or transformers where there are no metallic connections between the TN-C system and the public supply.

IT system

The supply is isolated from earth and therefore there is no shock or fire risk involved when an earth fault occurs. Protection is afforded by monitoring equipment which gives an audible warning if a fault occurs. This type of supply is used in mines, quarries and chemical processes where interruption of the process may create a hazardous situation. The system must not be connected to a public supply, and is shown in Figure 2.13.

Low voltage distribution in buildings

In domestic installations the final circuits for lights, sockets, cookers, immersion heating etc., are connected to separate ways in the consumer's unit mounted at the service position as shown in Figure 2.11.

In commercial or industrial installations a three phase 415 V supply must be distributed to appropriate equipment in addition to supplying single phase 240 V loads such as lighting. It is now common practice to establish industrial estates

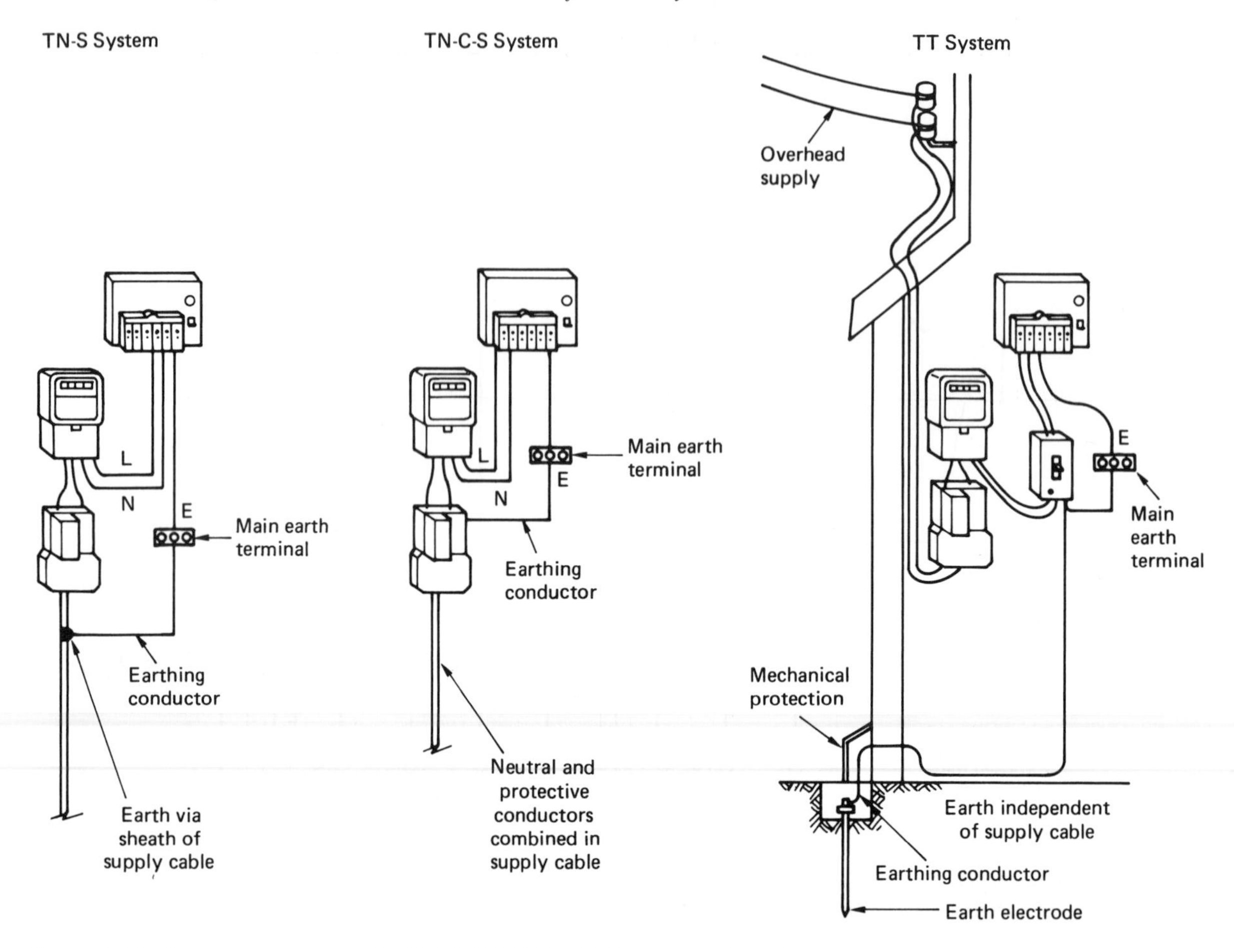

Figure 2.11 Service arrangements for TN-S, TN-C-S and TT systems of supply

speculatively, with the intention of encouraging local industry to use individual units. This presents the electrical contractor with an additional problem: The use and electrical demand of a single industrial unit is often unknown and the electrical supply equipment will need to be flexible in order to meet a changing demand due to expansion or change of use.

Busbar chambers incorporated into cubicle switchboards or on-site assemblies of switchboards are to be found at the incoming service position of commercial and industrial consumers, since this has proved to provide the flexibility required by these consumers. This is shown in Figure 2.14.

Distribution fuse boards, which may incorporate circuit breakers, are wired by sub-main cables from the service position to load centres in other parts of the building, thereby keeping the length of cable to the final circuit as short as possible. This is shown in Figure 2.15.

When high rise buildings such as multi-storey flats have to be wired, it is usual to provide a three phase four wire rising main. This may comprise vertical busbars running from top to bottom at some central point in the building. Each floor, individual flat or maisonette is then connected to the busbar to provide the consumer's supply. When individual dwellings receive a single phase supply the electrical contractor must balance the load across the three phases. Figure 2.16 shows a rising main system. The rising main must incorporate fire barriers to prevent the spread of fire throughout the building (Regulation 527–02).

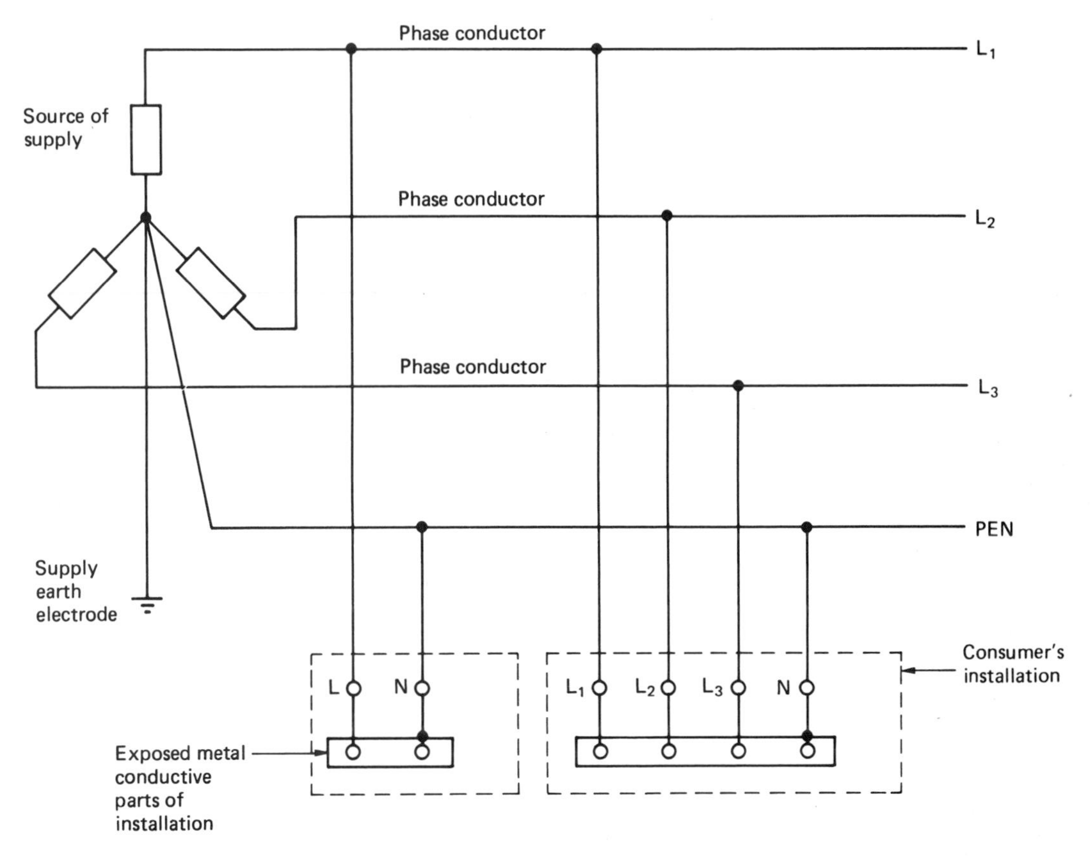

Figure 2.12 TN-C system – neutral and protective functions combined throughout system

Radial distribution calculations

A rising main is one example of a radial circuit with various loads connected along its radial length. The current carried by the rising main busbars close to the supply point will be greater than the current at the farthest point. The volt drop at the farthest point will be greatest when all loads are connected. The currents and volt drops in a radial distribution circuit may be calculated as illustrated by the following examples.

Example 1
A 240 V radial distributor is 90 m long and has three loads rated at 20 A, 40 A and 60 A connected at equal distances along the cable. The combined resistance of the live and neutral conductor is 0.5 mΩ per metre. Calculate the current in each section of the cable and the voltage at the end of the cable furthest away from the supply.

A simple line diagram as given by Figure 2.17 will help us to understand the circuit. The loads must be connected every 30 m.

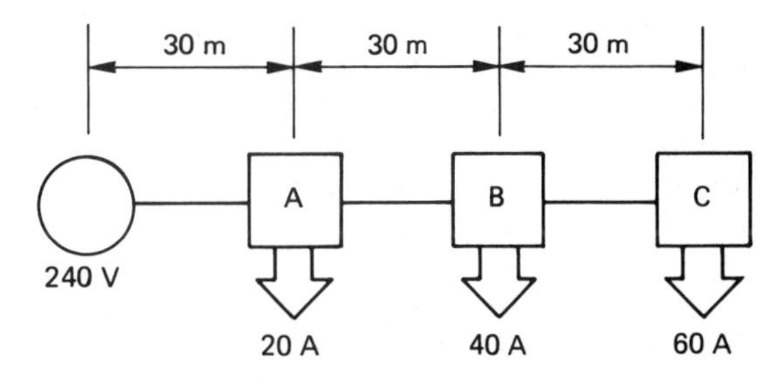

Figure 2.17 Simple line diagram for Example 1

The current in the cable between loads B-C
= 60 A.
The current in the cable between loads A-B
= 40 A + 60 A = 100 A.

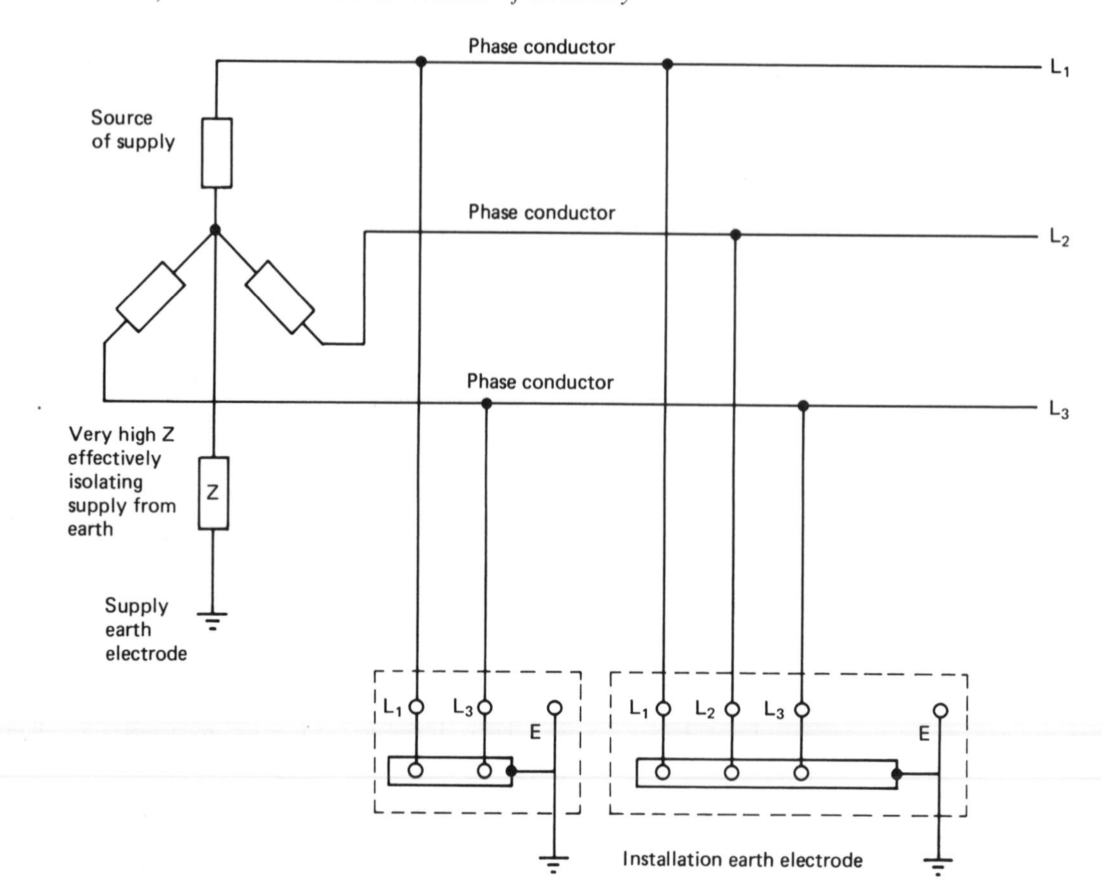

Figure 2.13 IT system – supply isolated from earth and installation's conductive parts connected to earth electrode

The current in the cable between the supply and load A = 20 A + 40 A + 60 A = 120 A.
The resistance of each section of the cable will be $30\ \text{m} \times 0.5 \times 10^{-3}\ \Omega/\text{m} = 0.015\ \Omega$.
The volt drop between the supply and load A will be

$$V = I \times R$$
$$\therefore \quad V = 120\ \text{A} \times 0.015\ \Omega = 1.8\ \text{V}.$$

The volt drop between load A and B will be

$$V = I \times R$$
$$\therefore \quad V = 100\ \text{A} \times 0.015\ \Omega = 1.5\ \text{V}.$$

The volt drop between load B and C will be

$$V = I \times R$$
$$\therefore \quad V = 60\ \text{A} \times 0.015\ \Omega = 0.9\ \text{V}.$$

The total volt drop = 1.8 + 1.5 + 0.9 = 4.2 V.

The total voltage drop of 4.2 V is acceptable since it is below the 9.6 V demanded by Regulation 525–01.

Voltage at the cable end = 240 V – 4.2 V
= 235.8 V.

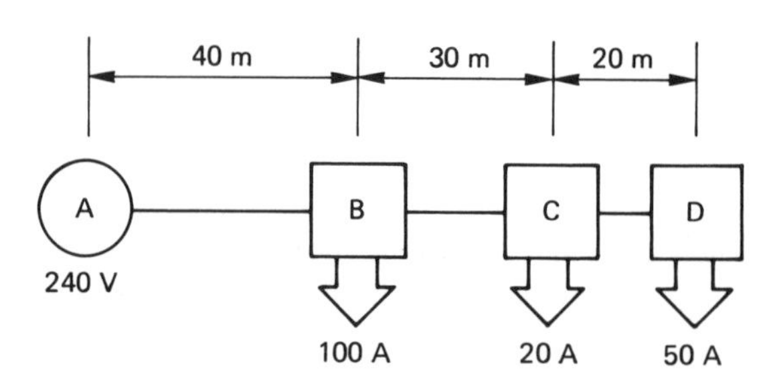

Figure 2.18 Simple line diagram for Example 2

Example 2
A 95 mm^2 two core PVC insulated armour cable with copper conductors is clipped direct to a surface and forms the radial distributor loaded as shown in Figure 2.18. Calculate the currents in each section of the cable. Using Appendix 4 of the IEE Regulations calculate the voltage at Point D.

The current in cable C-D = 50 A.
The current in cable B-C = 20 A + 50 A = 70 A.
The current in cable A-B = 100 A + 20 A + 50 A = 170 A.

From Appendix 4 of the IEE Regulations, Table 4D4B, the volt drop per ampere per metre for this cable is given as 0.5 mV, taking the impedance value.

The volt drop A-B $= 0.5 \times 10^{-3}\ \text{V} \times 170\ \text{A} \times 40\ \text{m} = 3.4\ \text{V}.$
The volt drop B-C $= 0.5 \times 10^{-3}\ \text{V} \times 70\ \text{A} \times 30\ \text{m} = 1.05\ \text{V}.$
The volt drop C-D $= 0.5 \times 10^{-3}\ \text{V} \times 50\ \text{A} \times 20\ \text{m} = 0.5\ \text{V}.$
The total volt drop $= 3.4\ \text{V} + 1.05\ \text{V} + 0.5\ \text{V} = 4.95\ \text{V}.$
The voltage at point D $= 240\ \text{V} - 4.95\ \text{V} = 235.05\ \text{V}.$

Ring distribution calculations

Ring distribution circuits form a closed loop and their solution therefore requires the assistance of the laws devised by Gustav Kirchhoff (1824–1887). Kirchhoff's first law states that the algebraic sum of the current at a junction is zero,

or $\Sigma I = 0$.

Kirchhoff's second law states that the algebraic sum of the emfs acting around a closed loop is equal to the algebraic sum of the voltage drop around the same loop,

or $\Sigma\ emf = \Sigma\ IR$.

Let us now apply these laws to the solution of a ring main problem.

Example
A ring main cable supplies loads B and C from a 240 V supply at point A as shown by Figure 2.19(a). The cable lengths are indicated on the diagram and the cable resistance (both live and neutral conductor) is 0.25 Ω per 1000 m. Find:

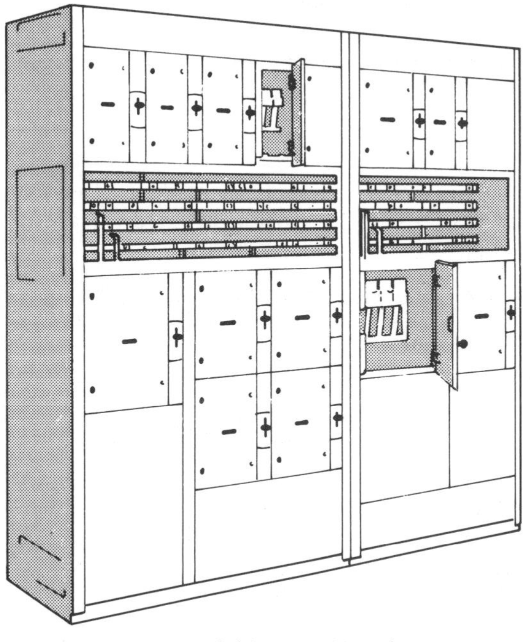

Cubicle switchboards

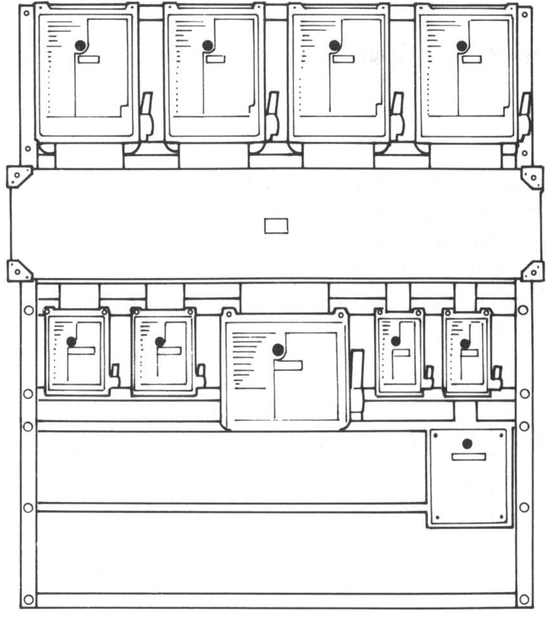

Switchboards

Figure 2.14 Industrial consumer's service position equipment

a) the current in each section of the ring main,
b) the voltage at loads B and C,
c) the supply current.

In finding a solution to this problem we must first make some statements about the circuit. For example, let us

- assume that all the current flows around the circuit in a clockwise direction,
- assume that the current in the cable A to B = I,
- assume that the current in the cable B to C = I − 70, that is the current I less the 70 A fed to load B,
- assume that the current in the cable C to A = I − 120, that is the current I less the 70 A fed to load B and the 50 A fed to load C.

These assumptions are shown in diagram (b) of Figure 2.19. The resistance of the cable sections

$$\text{A to B} = \frac{30\text{ m}}{1000\text{ m}} \times 0.25\ \Omega = 0.0075\ \Omega$$

$$\text{B to C} = \frac{40\text{ m}}{1000\text{ m}} \times 0.25\ \Omega = 0.01\ \Omega$$

$$\text{C to A} = \frac{90\text{ m}}{1000\text{ m}} \times 0.25\ \Omega = 0.0225\ \Omega.$$

By Kirchhoff's second law $\Sigma\ emf = \Sigma\ IR$. Now, the $\Sigma\ emf = 0$ for a closed loop supplied at one point and the ΣIR = (volt drop A to B) + (volt drop B to C) + (volt drop C to A). Substituting into Kirchhoff's second equation we have:

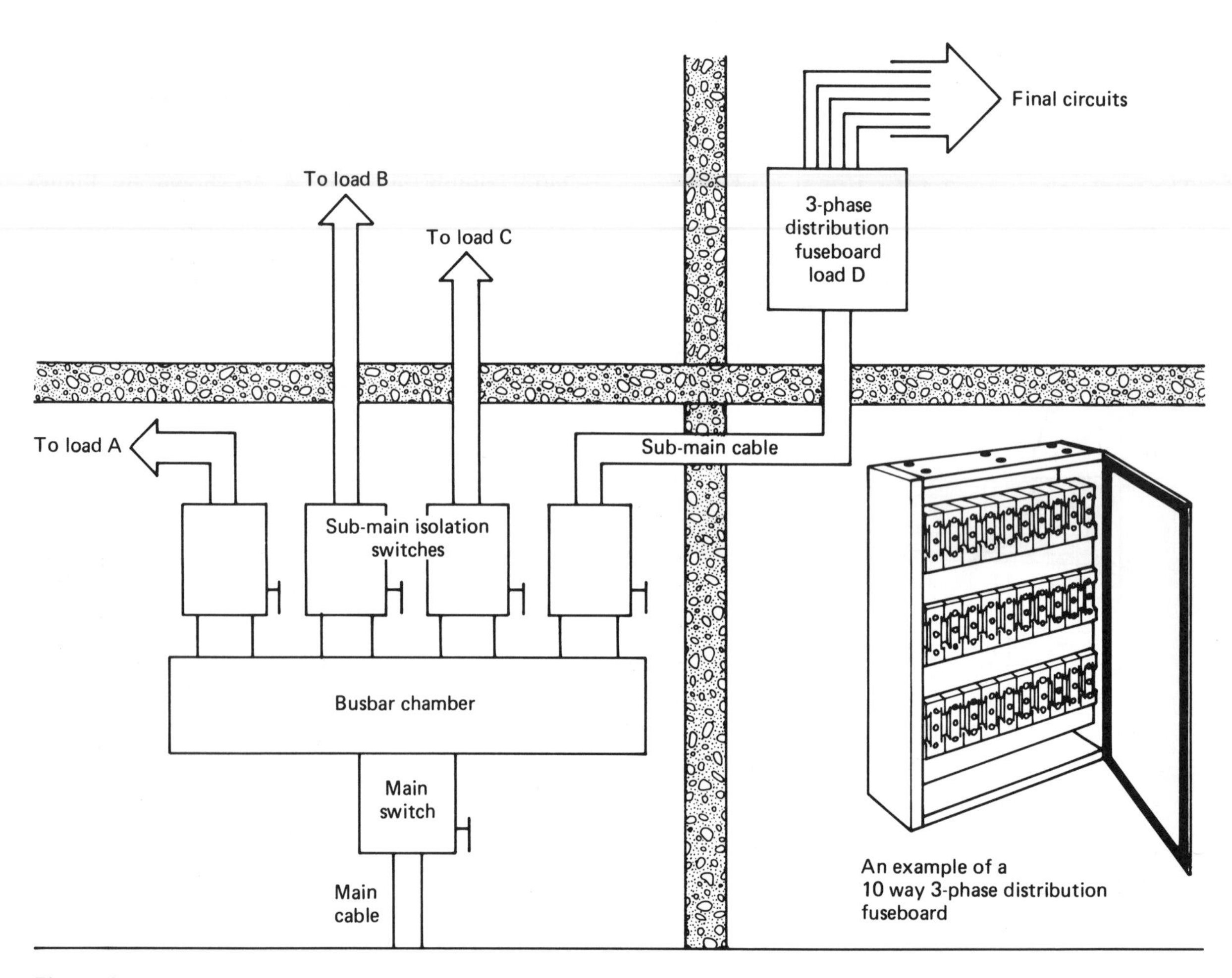

Figure 2.15 Typical distribution in commercial or industrial building

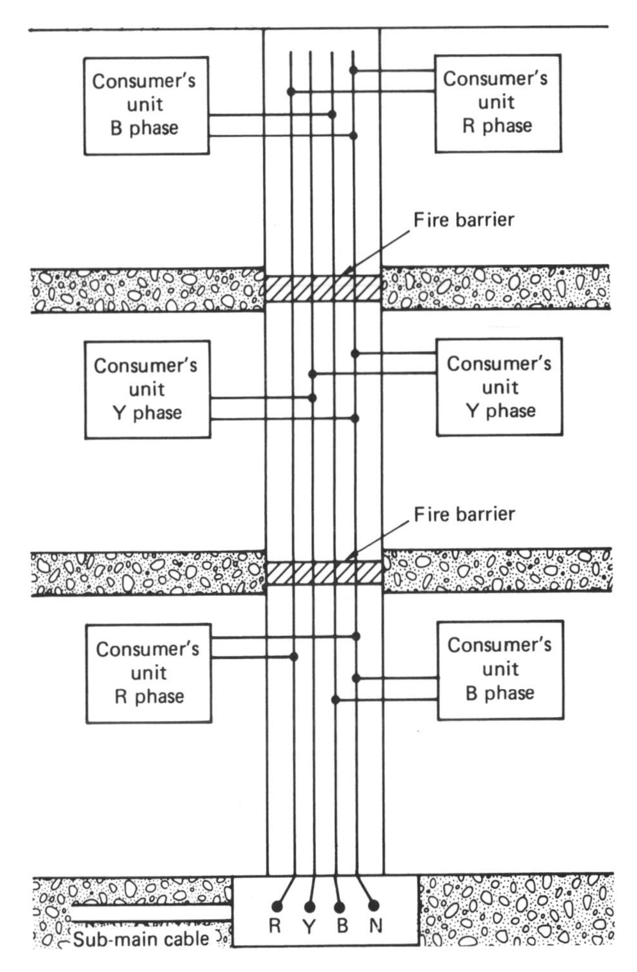

An example of a busbar chamber

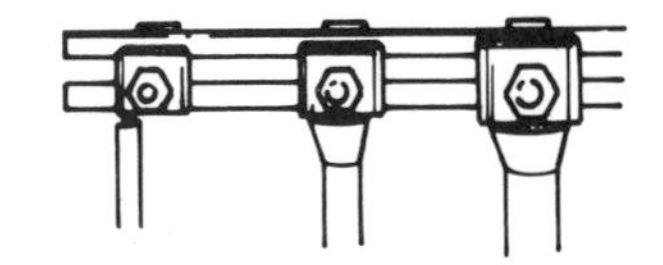

Consumer's supply cables are connected by clamps to the busbars

Figure 2.16 Busbar rising main system

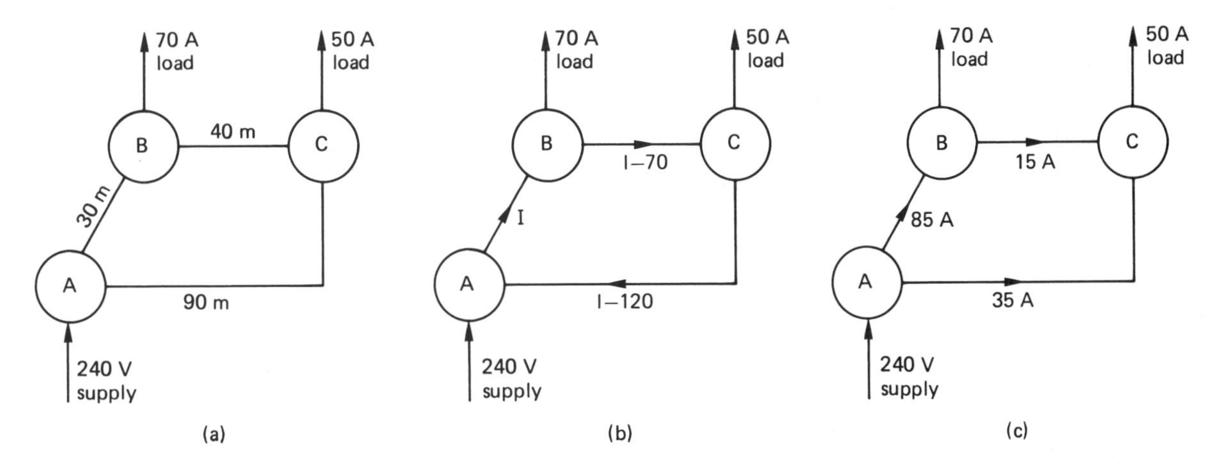

Figure 2.19 Line diagram for ring main calculation

$0 = (I \times 0.0075) + [(I - 70) \times 0.01] + [(I - 120) \times 0.0225]$
$0 = 0.0075\,I + 0.01\,I - 0.7 + 0.0225\,I - 2.7$
$0 = 0.04\,I - 3.4$ (by collecting terms)

$$\therefore \quad I = \frac{3.4}{0.04} = 85\text{ A}.$$

From Figure 2.19(b):

The current in the cable A to B = I = 85 A.
The current in the cable B to C = $I - 70$ = 85 A − 70 A = 15 A.
The current in the cable C to A = $I - 120$ = 85 A − 120 A = − 35 A.

The negative sign indicates that the current flows from A to C and not from C to A as we assumed. Therefore the 50 A load at C is supplied by 15 A via cable BC and 35 A via cable AC as shown by diagram (c) in Figure 2.19.

The voltage at load B = voltage at A − volt drop A to B

$\therefore$ voltage at load B = 240 V − (85 A × 0.0075 Ω)
= 240 V − 0.6375 V
= 239.36 V.

The voltage at load C = voltage at A − volt drop A to C

$\therefore$ voltage at load C = 240 V − (35 A × 0.0225 Ω)
= 240 V − 0.788 V
= 239.2 V.

From Kirchhoff's first law,

$\Sigma I = 0$.

Assuming currents going into a circuit to be positive and those going out of a circuit to be negative we have:

$$\text{Supply current} - \text{Load current B} - \text{Load current C} = 0$$

$$\text{Supply current} = \text{Load current B} + \text{Load current C}$$

$\therefore$ Supply current = 70 A + 50 A = 120 A.

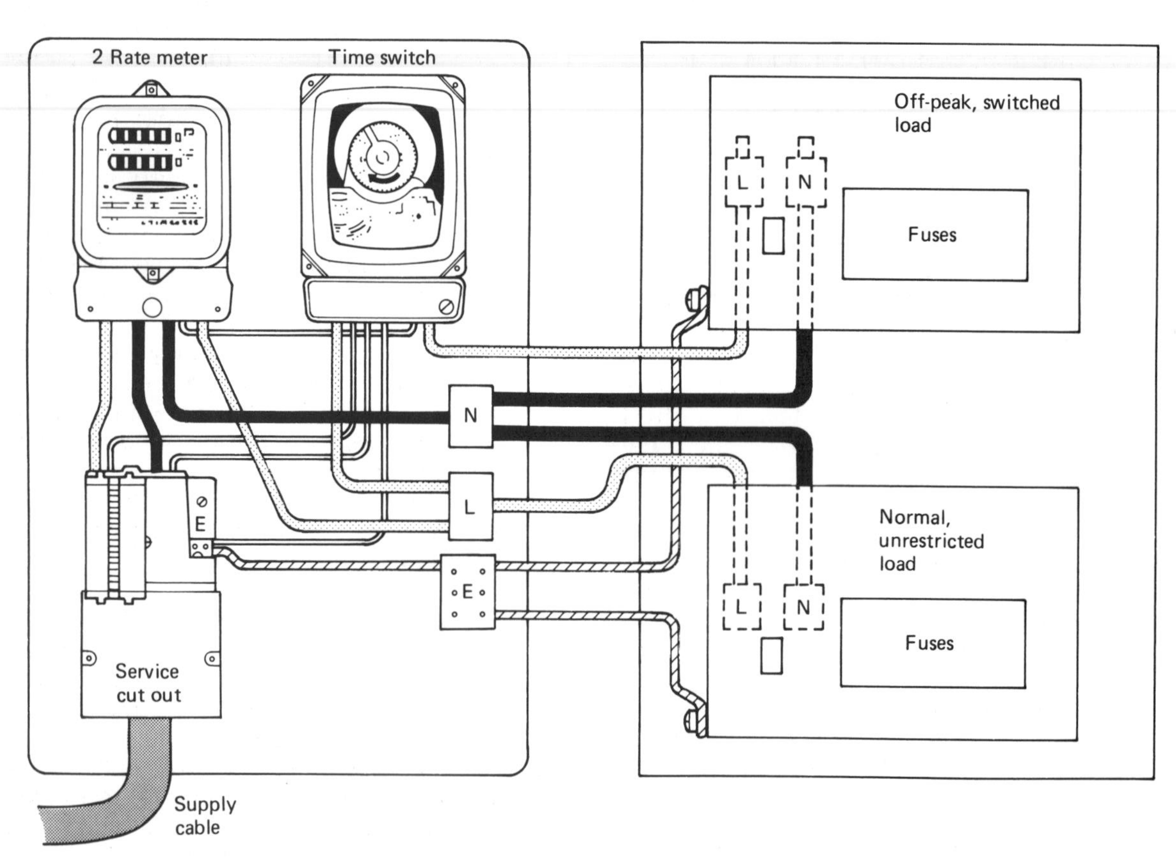

Figure 2.20 Service position equipment required for an off-peak supply

Off-peak supplies

Off-peak supplies are those made available by supply authorities to consumers during restricted periods, usually during the night. They attract a lower tariff than unrestricted supplies. (The off-peak charge is usually about half the normal supply charge). Recent years have seen a growing demand for off-peak supplies by domestic consumers, partly as a result of the publicity given to the 'White meter' and the introduction of new materials making block storage heaters more efficient, slim and attractive.

The White meter is a two rate meter. Units consumed during the day and evening are charged at the normal rate and all units consumed during the night are charged at a lower rate. To take advantage of an off-peak supply, the consumer must install storage heaters and heat the domestic hot water during the night, keeping it hot during the day by improved levels of thermal insulation.

The consumer's equipment at the service position is divided into two parts; an unrestricted supply feeding lighting, cooking and socket outlet final circuits, and a restricted supply, controlled by the supply authority's time switch, feeding space heating and water heating final circuits. The service position arrangements for a domestic consumer are shown in Figure 2.20. The supply authority's equipment and connector blocks are sealed to prevent unauthorised entry.

Exercises

1 The transmission of electricity is, for the most part, by overhead conductors suspended on steel towers because:
(a) this is environmentally more acceptable than running cables underground, (b) this is very many times cheaper than running an equivalent cable underground, (c) high voltage electricity cables cannot be buried on agricultural land, (d) more power can be carried by overhead conductors than is possible by an underground cable.

2 A ring distribution system of electrical supply:
(a) is cheaper than a radial distribution system, (b) can use smaller supply cables than a radial distribution system, (c) offers greater security of supply than a radial distribution system, (d) is safer than a radial system.

3 A TN-S system of electrical supply:
(a) uses a separate neutral and protective conductor throughout the system, (b) uses a combined protective and neutral conductor, (c) does not provide the consumer with an earth terminal and the installations circuit protective conductors must be connected to earth by an earth electrode provided by the consumer, (d) is isolated from earth.

4 A TN-C-S system of electrical supply:
(a) uses a separate neutral and protective conductor throughout the system, (b) uses a combined protective and neutral conductor, (c) does not provide the consumer with an earth terminal and the installations circuit protective conductors must be connected to earth by an earth electrode provided by the consumer, (d) is isolated from earth.

5 An off-peak supply is available to a consumer:
(a) for water heating purposes only, (b) during the day time only, (c) during the night time only, (d) at all times provided that peak demand does not exceed 4 hours.

6 Sketch an 11 kV ring distribution system suitable for supplying four load centres in a factory from a 33 kV primary sub-station. Show how the supply can be secured to all four load centres even though a fault occurs on a cable between two of the distribution sub-stations.

7 Use a block diagram to describe the electrical equipment which would be installed in an 11 kV/415 V sub-station of brick construction.

8 Use a simple circuit diagram to show how a 240 V and 415 V consumer may be connected to the supply authority's TN-S system. Include in your diagram the supply authority's transformer.

9 Using a circuit diagram which shows the secondary winding of the supply transformer, show how a 240 V and 415 V supply are connected to a TN-C-S system.

10 Show how a 240 V and 415 V installation is

connected to a TT system of supply. Clearly explain how an effective earth is obtained for the installation and the possible dangers in not providing an adequate earth connection.

11 Sketch the arrangements of the service position equipment at a 240 V installation supplied by
(a) a TN-S system of supply, (b) a TN-C-S system of supply, (c) a TT system of supply.

12 A 415 V three phase and neutral busbar rising main is to be used to provide a 240 V supply to each of 8 individual flats on 4 floors of a building.
(a) Sketch the arrangement and describe how each flat supply must be connected to the rising main if the total load is to be balanced.
(b) Describe the method used to prevent the spread of fire.

13 State the advantages and disadvantages of transmitting electricity
(a) at very high voltage, (b) by overhead lines suspended on steel towers.

14 (a) State the meaning of an off-peak supply.
(b) An installation is made up of a lighting load, a cooking load, a space heating load and a water heating load. Which of these loads are suitable for connection to an off-peak supply? Sketch the arrangements required at the service position to make the connections.

15 A 90 m radial distributor is equally divided into three sections AB, BC and CD each 30 m long. A 240 V supply is connected at point A, a load of 20 A is connected at point B, a load of 30 A is connected at point C and a load of 40 A at point D.
(a) Sketch the arrangement. (b) Calculate the current in each section of the radial distributor. (c) Calculate the voltage at point D if the cable resistance is 0.5 mΩ per metre.

16 A 70 mm^2 two core armoured PVC insulated cable with copper conductors is clipped direct to a surface and forms the radial distributor loaded as shown in Figure 2.21.

Figure 2.21 Line diagram for Question 16

The distributor is fed at end A with 240 V and the cable lengths are: A to B 50 m, B to C 30 m, C to D 20 m.
(a) Sketch the diagram and show the currents in each section of the distributor.
(b) Using Appendix 4 of the IEE Regulations, calculate the voltage of the load points B, C and D.

CHAPTER 3

Protection, isolation and switching of consumers' equipment

Every year there are approximately 50 fatal electrocutions involving electrical equipment and some 20 000 fires attributed to electrical causes. The majority of accidents are due to carelessness or misuse of electrical equipment or repairs attempted without first disconnecting the circuit from the source of supply. In recent years there has also been a trend toward loading electrical installations to the limit of their capacity for economic reasons, which has also reduced safety margins.

An electrical installation which is designed and installed in accordance with the appropriate regulations will give good protection against an electric shock and the risk of fire, but the ultimate safety of the individual depends upon the good sense of the user of the electrical equipment.

Part 4 of the IEE Regulations deals with the application of protective measures for safety and Chapter 53 with the regulations for switching devices or switchgear required for protection, isolation and switching of a consumer's installation.

The consumer's main switchgear must be readily accessible to the consumer and be able to:

- isolate the complete installation from the supply,
- protect against overcurrent and
- cut off the current in the event of a serious fault occurring.

Isolation and switching

The Regulations identify four separate types of switching; switching for isolation, mechanical maintenance, emergency switching and functional switching.

Isolation is defined as cutting off the electrical supply to a circuit or item of equipment in order to ensure the safety of those working on the equipment by making dead those parts which are live in normal service.

An isolator is a mechanical device which is operated manually and used to open or close a circuit off load. An isolator switch must be provided close to the supply point so that all equipment can be made safe for maintenance. Isolators for motor circuits must isolate the motor and the control equipment, and isolators for high voltage discharge lighting luminaires must be an integral part of the luminaire so that it is isolated when the cover is removed (Regulations 461, 476–02 and 537–02). Devices which are suitable for isolation are isolation switches, fuse links, circuit breakers, plugs and socket outlets.

Isolation at the consumer's service position can be achieved by a double pole switch which opens or closes all conductors simultaneously. On three phase supplies the switch need only break the live conductors with a solid link in the neutral, provided that the neutral link cannot be removed before opening the switch.

The switching for mechanical maintenance requirements is similar to those for isolation except that the control switch must be capable of switching the full load current of the circuit or piece of equipment. Switches for mechanical maintenance must not have exposed live parts when the appliance is opened, must be connected in the main electrical circuit and have a reliable on/off indication or visible contact gap (Regulations 462 and 537–03). Devices which are suitable for switching off for mechanical maintenance are switches, circuit breakers, plug and socket outlets.

Emergency switching involves the rapid disconnection of the electrical supply by a single action to remove or prevent danger. The device used for emergency switching must be immediately accessible and identifiable, and be capable of cutting off the full load current. A fireman's switch provides emergency switching for high voltage signs as described in Chapter 9.

Electrical machines must be provided with a means of emergency switching, and a person operating an electrically driven machine must have access to an emergency switch so that the machine can be stopped in an emergency. The remote stop/start arrangement shown in Figure 7.10 could meet this requirement for an electrically driven machine (Regulations 463, 476–03 and 537-04). Devices which are suitable for emergency switching are switches, circuit breakers and contactors. Where contactors are operated by remote control they should *open* when the coil is de-energised, that is, fail safe. Push buttons used for emergency switching must be coloured red and latch in the stop or off position. They should be installed where danger may arise and be clearly identified as emergency switches. Plugs and socket outlets cannot be considered appropriate for emergency disconnection of supplies.

Functional switching involves the switching of electrically operated equipment in normal service. The device must be capable of interrupting the total steady current of the circuit or appliance. When the device controls a discharge lighting circuit it must have a current rating capable of switching an inductive load. Plug and socket outlets may be used as switching devices and recent years have seen an increase in the number of electronic dimmer switches being used for the control and functional switching of lighting circuits (Regulations 537–05–01).

Where more than one of these functions is performed by a common device, it must meet the individual requirements for each function (Regulation 476–01–01).

Overcurrent protection

The consumer's mains equipment must provide protection against overcurrent (Regulation 431). Fuses provide overcurrent protection when situated in the live conductors; they must not be connected in the neutral conductor. Circuit breakers may be used in place of fuses, in which case the circuit breaker may also provide the means of isolation, although a further means of isolation is usually provided so that maintenance can be carried out on the circuit breakers themselves.

The term overcurrent can be sub-divided into overload current, and short circuit current. An overload current can be defined as a current which exceeds the rated value in an otherwise healthy circuit. Overload currents usually occur because the circuit is abused or because it has been badly designed or modified. A short circuit is an overcurrent resulting from a fault of negligible impedance connected between conductors. Short circuits usually occur as a result of an accident which could not have been predicted before the event.

An overload may result in currents of two or three times the rated current flowing in the circuit. Short circuit currents may be hundreds of times greater than the rated current. In both cases the basic requirements for protection are that the fault currents should be interrupted quickly and the circuit isolated safely before the fault current causes a temperature rise which might damage the insulation and terminations of the circuit conductors.

The selected protective device should have a current rating which is not less than the full load current of the circuit but which does not exceed the cable current rating. The cable is then fully protected against both overload and short circuit faults. (Regulation 433–02–01). Devices which provide over-current protection are:

- HBC fuses to BS 88. These are high rupturing capacity fuses for industrial applications having a maximum fault capacity of 80 kA.
- Cartridge fuses to BS 1361. These are used for a.c. circuits on industrial and domestic installations having a fault capacity of about 50 kA.
- Cartridge fuses to BS 1362. These are used in 13 A plug tops and have a maximum fault capacity of about 6 kA.
- Semi-enclosed fuses to BS 3036. These were previously called rewireable fuses and are used mainly on domestic installations having a maximum fault capacity of about 2 kA.
- MCBs to BS 3871. These are miniature circuit breakers which may be used as an alternative

to fuses for some installations. The British Standard includes ratings up to 100 A and maximum fault capacities of 9 kA. They are graded in four categories according to their instantaneous tripping currents; that is the current at which they will trip within 100 milliseconds.

Type 1 MCB to BS 3871 will trip instantly at between 2.7 and four times its rated current and is therefore more suitable on loads with little or no switching surges such as domestic installations.

Type 2 MCB to BS 3871 will trip instantly at between four and seven times its rated current. It offers fast protection on small overloads combined with a slower operation on heavier faults which reduces the possibility of nuisance tripping. The characteristics are very similar to an HBC fuse and this MCB is possibly best suited for general industrial use.

Type 3 MCB to BS 3871 will trip instantly at between seven and ten times its rated current. It is more suitable for protecting highly inductive circuits and is used on circuits supplying transformers chokes and lighting banks.

Type 4 MCB to BS 3871 will trip instantly between 10 and 50 times the rated current and is more suitable for special industrial applications such as welding equipment and X-ray machines.

The construction, advantages and disadvantages of the various protective devices are discussed in Chapter 8 of *Basic Electrical Installation Work*.

Position of protective devices

The general principle to be followed is that a protective device must be placed at a point where a reduction occurs in the current carrying capacity of the circuit conductors. A reduction may occur because of a change in the size or type of conductor or because of a change in the method of installation or a change in the environmental conditions. The only exceptions to this rule are where an overload protective device opening a circuit might cause a greater danger than the overload itself. For example, a circuit feeding an overhead electromagnet in a scrapyard.

Disconnection time calculations

The overcurrent protection device protecting socket outlet circuits and any fixed equipment in bathrooms must operate within 0.4 seconds. Those protecting fixed equipment circuits in rooms other than bathrooms must operate within five seconds (Regulation 413–02–08 to 13 and 601).

The reason for the more rapid disconnection of the socket outlet circuits is that portable equipment plugged into the socket outlet is considered a higher risk than fixed equipment since it is more likely to be firmly held by a person. The more rapid disconnection times for fixed equipment in bathrooms take account of a possibly reduced body resistance in the bathroom environment.

The IEE Regulations permit us to assume that where an overload protective device is also intended to provide short circuit protection, and has a rated breaking capacity greater than the prospective short circuit current at the point of its installation, the conductors on the load side of the protective device are considered to be adequately protected against short circuit currents without further proof. This is because the cable rating and the overload rating of the device are compatible. However, if this condition is not met or if there is some doubt, it must be verified that fault currents will be interrupted quickly before they can cause a dangerously high temperature rise in the circuit conductors. Regulation 434–03–03 provides an adiabatic equation for calculating the maximum operating time of the protective device to prevent the permitted conductor temperature rise being exceeded as follows:

$$t = \frac{k^2 S^2}{I^2} \text{ (seconds)}$$

where t = duration time in second
S = cross-sectional area of conductor in mm^2
I = short circuit rms current in amperes
k = a constant dependent upon the conductor metal and type of insulation. Values are given in Table 43A.

Example

A 10 mm PVC insulated copper cable is short circuited when connected to a 415 V supply. The impedance of the short circuit path is 0.1 Ω. Calculate the maximum permissible disconnection time and show that a 50 A type 2 MCB to BS 3871 will meet this requirement.

$$I = \frac{V}{Z} \text{ (A)} \qquad \therefore \quad I = \frac{415 \text{ V}}{0.1\ \Omega} = 4150 \text{ A}.$$

k for copper conductor and PVC insulation is 115 from Table 43A of the IEE Regulations.

$$t = \frac{k^2\ S^2}{I^2}\ (\text{s})$$

$$\therefore \quad t = \frac{115^2 \times 10^2\ \text{mm}^2}{4150^2\ \text{A}} = 76.79 \times 10^{-3}\ \text{s}.$$

The maximum time that a 4150 A fault current can be applied to this 10 mm^2 cable without dangerously raising the conductor temperature is 0.07679 seconds. Therefore, the protective device must disconnect the supply to the cable in less than 0.07679 seconds under short circuit conditions. Manufacturers' information and Appendix 3 of the IEE Regulations give the operating times of protective devices at various short circuit currents.

Time/current characteristics of protective devices

Disconnection times for various overcurrent devices are given in the form of a logarithmic graph. This means that each successive graduation of the axis represents a ten times change of the previous graduation.

These logarithmic scales are shown in the graphs of Figures 3.1 and 3.2. From Figure 3.1 it can be seen that the particular protective device will take 8 seconds to disconnect a fault current of 50 A and 0.08 seconds to clear a fault current of 1000 A.

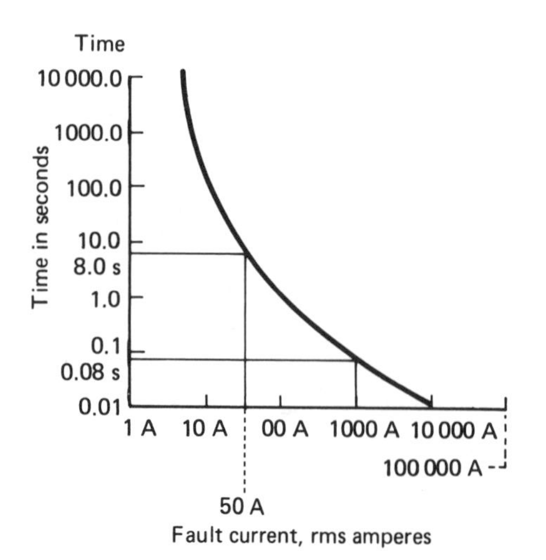

Figure 3.1 Time/current characteristic of an overcurrent protective device

Figure 3.2 shows the time/current characteristics for a type 2 miniature circuit breaker to BS 3871. This is the type of protective device used in Example 1. This graph shows that a fault current of 4150 A will trip the protective device in 0.02 seconds. Since this is quicker than 0.07679 seconds, the 50 A type 2 MCB will clear the fault current before the temperature of the cable is raised to a dangerous level.

Appendix 3 of the IEE Regulations gives the time/current characteristics and specific values of prospective short circuit current for a number of protective devices.

These indicate the value of fault current which will cause the protective device to operate in the times indicated by Chapter 413 of the IEE Regulations, that is 0.4 and 5 seconds in the case of domestic socket outlet circuits and distribution circuits feeding fixed appliances.

Figures 1, 2 and 3 in Appendix Three deal with fuses and Figures 4 to 8 with miniature circuit breakers.

It can be seen that the prospective fault current required to trip an MCB in the required time is a multiple of the current rating of the device. The multiple depends upon the characteristics of the particular devices. Thus:

Type 1 MCB to BS 3871 has a multiple of 4
Type 2 MCB to BS 3871 has a multiple of 7
Type 3 MCB to BS 3871 has a multiple of 10
Type B MCB to BS 3871 has a multiple of 5
Type C MCB to BS 3871 has a multiple of 10

Example 1:

A 6 A type 1 MCB to BS 3871 used to protect a domestic lighting circuit will trip within 5 seconds when 6 A times a multiple of 4, that is 24 A flows under fault conditions.

Therefore if the earth fault loop impedance is low enough to allow at least 24 A to flow in the circuit under fault conditions, the protective device will operate within the time required by Regulation 413–02–14. Therefore the maximum value of earth fault loop impedance for a domestic lighting circuit protected with a 6 A type 1 MCB will be:

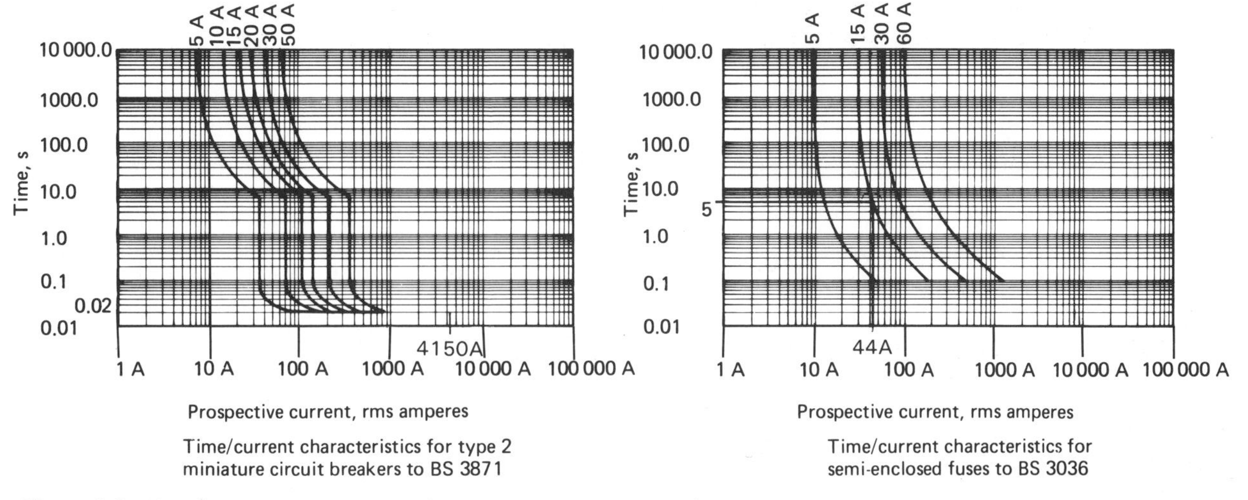

Figure 3.2 Time/current characteristics for overcurrent protective devices

$$Z_S = \frac{V}{1}\ \Omega$$

$$\therefore\quad Z_S = \frac{240\ V}{24\ A} = 10\ \Omega$$

The maximum permitted value of earth loop impedance is therefore 10 Ω which is the value given in line E of Table 41B2.

Example 2. A 32A MCB to BS 3871 is to be used to protect a 240V industrial socket outlet circuit. Find the maximum value of earth fault loop impedance to allow the protective device to operate within 0.4 seconds under fault conditions.

Minimum tripping current for this device is 32A times a multiple of 7, that is 224A. If the earth fault loop impedance is low enough to allow 224A to flow under fault conditions then the conditions of Regulation 413–02–09 will be met. The maximum value of earth loop impedance is given by

$$Z_S = \frac{V}{I}\ \Omega$$

$$\therefore\quad Z_S = \frac{240\ V}{224\ A} = 1.07\ \Omega$$

The maximum permissible value of earth loop impedance is therefore 1.07 Ω which is the value given in line F of Table 41B2.

The characteristics of Appendix 3 give the specific values of prospective short circuit current for all standard sizes of protective device.

Earth fault loop impedance Z_s

In order that an overcurrent protective device can operate successfully, meeting the required disconnection times, the earth fault loop impedance must be less than those values given in Table 41B1 and 41B2 for socket outlet circuits and Table 41B2 and 41D for circuits supplying fixed equipment. The value of the earth fault loop impedance may be verified by means of an earth fault loop impedance test as described in Chapter 9 of *Basic Electrical Installation Work*, or calculated as follows:

$$Z_S = Z_E + (R_1 + R_2)\ (\Omega)$$

where Z_E = the impedance of the supply side of the earth fault loop. The actual value will depend upon many factors; the type of supply, the ground conditions, the distance from the transformer etc.. The value can be obtained from the Area Electricity Board, but typical values are as follows:

$Z_E = 0.35\ \Omega$ for TN-C-S (PME) supplies
$Z_E = 0.8\ \Omega$ for TN-S (cable sheath earth) supplies
R_1 = the resistance of the phase conductor
R_2 = the resistance of the earth conductor.

The complete earth fault loop path is shown in Figure 3.3.

Values of $(R_1 + R_2)$ have been calculated for copper and aluminium conductors and are given in Table 6A of the On Site Guide.

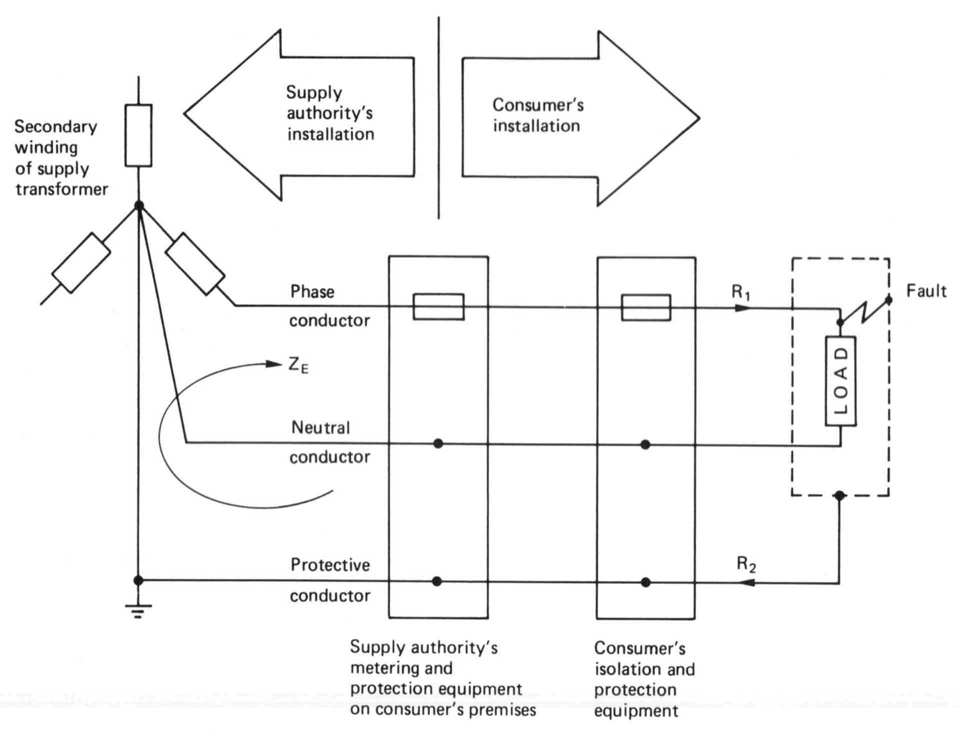

Figure 3.3 Earth fault loop path for a TN-S system

Example
A 20 A radial socket outlet circuit is wired in $2.5\,\text{mm}^2$ PVC cable incorporating a $1.5\,\text{mm}^2$ CPC. The cable length is 30 m and the consumer's protection is by semi-enclosed fuse to BS 3036. The earth fault loop impedance of the supply is 0.5 Ω. Calculate the total earth fault loop impedance Z_S, and establish that the value is less than the maximum value permissible for this type of circuit.

$Z_S = Z_E + (R_1 + R_2)$ (Ω)
$Z_E = 0.5$ Ω (value given in the question).

From the value given in Table 6A,

$(R_1 + R_2) = 19.51 \times 10^{-3}\ \Omega/\text{m} \times 30\ \text{m} = 0.585\ \Omega$

However, under fault conditions, the temperature and therefore the cable resistance will increase. To take account of this, we must multiply the value of cable resistance by the factor given in Table 6B of the On Site Guide.

In this case the factor is 1.38 and therefore the cable resistance under fault conditions will be:

$0.585\ \Omega \times 1.38 = 0.8073\ \Omega$

The total earth fault loop impedance is therefore

$Z_S = 0.5\ \Omega + 0.8073\ \Omega = 1.307\ \Omega$

The maximum permitted value given in Table 41B1 for a 20 A fuse to BS 3036 is 1.85 Ω. The circuit earth fault loop impedance is less than this value and therefore the protective device will operate within the required disconnection time of 0.4 s.

Protective conductor size

The circuit protective conductor forms an integral part of the total earth fault loop impedance, so it is necessary to check that the cross-section of this conductor is adequate. If the cross-section of the circuit protective conductor complies with Table 54G of the IEE Regulations, there is no need to carry out further checks. Where phase and protective conductors are made from the same material Table 54G tells us that:

1. for phase conductors equal to or less than

16 mm^2, the protective conductor should equal the phase conductor

2. for phase conductors greater than 16 mm^2 but less than 35 mm^2, the protective conductor should have a c.s.a. of 16 mm^2
3. for phase conductors greater than 35 mm^2, the protective conductor should be half the size of the phase conductor.

However, where the conductor cross-section does not comply with this table, then the formula given in Regulation 543–01–03 must be used.

$$S = \frac{\sqrt{I^2 t}}{k} \text{ (mm}^2\text{)}$$

where S = cross sectional area in mm^2
I = value of maximum fault current in amperes
t = operating time of the protective device
k = a factor for the particular protective conductor. Values are given in Tables 54B to 54F of the IEE Regulations.

Example 1

A 240 V ring main circuit of socket outlets is wired in 2.5 mm single PVC copper cables in a plastic conduit with a separate 1.5 mm CPC. An earth fault loop impedance test identifies Zs as 1.15 Ω. Verify that the 1.5 mm CPC meets the requirements of Regulation 543–01–03 when the protective device is a 30 A semi-enclosed fuse.

$$I = \text{Maximum fault current} = \frac{V}{Z_S} \text{ (A)}$$

$$\therefore I = \frac{240}{1.15} = 208.7 \text{ A}$$

t = Maximum operating time of the protective device for a socket outlet circuit is 0.4 seconds from Regulation 413–02–09. From Figure 3.2 you can see that the time taken to clear a fault of 208.7 A is about 0.4 seconds.

k = 115 (from Table 54C).

$$S = \frac{\sqrt{I^2 t}}{k} \text{ (mm}^2\text{)}$$

$$S = \frac{\sqrt{(208.7)^2 \text{ A} \times 0.4 \text{ s}}}{115} = 1.15 \text{ mm}^2.$$

A 1.5 mm^2 CPC is acceptable since this is the nearest standard size conductor above the minimum cross sectional area of 1.15 mm^2 found by calculation.

Example 2

A domestic immersion heater is wired in 2.5 mm^2 PVC insulated copper cable and incorporates a 1.5 mm^2 CPC. The circuit is correctly protected with a 15 A semi-enclosed fuse to BS 3036. Establish by calculation that the CPC is of an adequate size to meet the requirements of Regulation 543–01–03. The characteristics of the protective device are given in Figure 3.2.

For circuits feeding fixed appliances the maximum operating time of the protective device is 5 seconds. From Figure 3.2 it can be seen that a current of about 44 A will trip the 15 A fuse in 5 seconds. Alternatively Table 2A in Appendix 3 of the Regulations gives a value of 43A. Let us assume a value of 43A.

$\therefore$ I = 43 A
t = 5 seconds for fixed appliances
k = 115 (from Table 54C)

$$S = \frac{\sqrt{I^2 t}}{k} \text{ (mm}^2\text{) (from Regulation 543–01–03)}$$

$$S = \frac{\sqrt{(43)^2 \text{ A} \times 5 \text{ s}}}{115} = 0.86 \text{ mm}^2$$

The circuit protective conductor of the cable is greater than 0.86 mm^2 and is therefore suitable. If the protective conductor is a separate conductor, that is, it does not form part of a cable as in Example 2 and is not enclosed in a wiring system as in Example 1, the cross-section of the protective conductor must be not less than 2.5 mm^2 where mechanical protection is provided or 4.0 mm^2 where mechanical protection is *not* provided in order to comply with Regulation 547–03–03.

Discrimination

In the event of a fault occurring on an electrical installation only the protective device nearest to the fault should operate, leaving other healthy circuits unaffected. A circuit designed in this way would be considered to have effective discrimination. Effective discrimination can be achieved by graded protection since the speed of operation

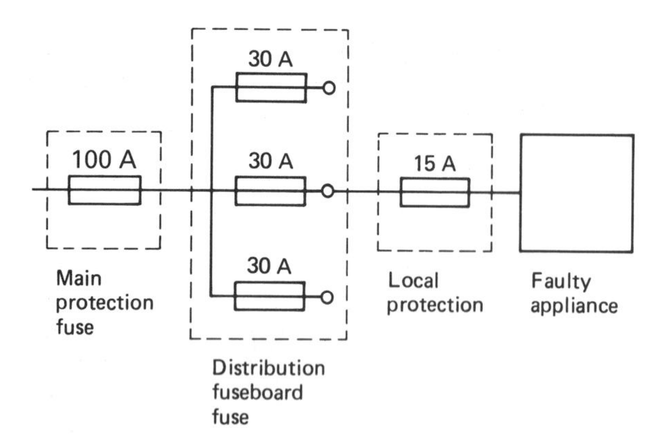

Figure 3.4 Effective discrimination achieved by graded protection

of the protective device increases as the rating decreases. This can be seen in Figure 3.2. A fault current of 200 A will cause a 15 A semi-enclosed fuse to operate in about 0.1 seconds, a 30 A semi-enclosed fuse in about 0.45 seconds and a 60 A semi-enclosed fuse in about 5.4 seconds. If a circuit is arranged as shown in Figure 3.4 and a fault occurs on the appliance, effective discrimination will be achieved because the 15 A fuse will operate more quickly than the other protective devices if they were all semi-enclosed types fuses with the characteristics shown in Figure 3.2.

Security of supply, and therefore effective discrimination, is an important consideration for an electrical engineer and is also a requirement of Regulation 533–01–06.

Exercises

1 The Regulations define isolation switching as:
(a) a mechanical switching device capable of making, carrying and breaking current under normal circuit conditions, (b) cutting off an electrical installation or circuit from every source of electrical energy, (c) the rapid disconnection of the electrical supply to remove or prevent danger, (d) the switching of electrical equipment in normal service.

2 Functional switching may be defined as:
(a) a mechanical switching device capable of making carrying and breaking current under normal circuit conditions, (b) cutting off an electrical installation or circuit from every source of electrical energy, (c) the rapid disconnection of the electrical supply to remove or prevent danger, (d) the switching of electrical equipment in normal service.

3 Emergency switching can be defined as:
(a) a mechanical switching device capable of making, carrying and breaking current under normal circuit conditions, (b) cutting off an electrical installation or circuit from every source of electrical energy, (c) the rapid disconnection of the electrical supply to remove or prevent danger, (d) the switching of electrical equipment in normal service.

4 The Regulations require that an overcurrent protective device interrupts a fault quickly and isolates the circuit before:
(a) the voltage on any extraneous conductive parts reaches 50 V, (b) the earth loop impedance reaches 0.4 Ω on circuits feeding 13 A socket outlets, (c) the fault causes damage to the circuit isolating switches, (d) the fault causes a temperature rise which might damage the insulation and terminations of the circuit conductors.

5 The maximum permissible value of the earth loop impedance of a circuit supplying fixed equipment and protected by a 30 A semi-enclosed fuse to BS 3036 is found by reference to the Tables in Part 4 to be:
(a) 1.1 Ω, (b) 1.92 Ω, (c) 2.0 Ω, (d) 2.76 Ω.

6 The earth fault loop impedance of a socket outlet circuit protected by a 30 A cartridge fuse to BS 1361 must not exceed:
(a) 0.4 Ω, (b) 1.14 Ω, (c) 1.20 Ω, (d) 2.0 Ω

7 The $(R_1 + R_2)$ resistance of 1000 m of PVC insulated copper cable having a 4.0 mm^2 phase conductor and 2.5 mm^2 protective conductor will be found from Table 6A to be
(a) 4.61 Ω, (b) 9.22 Ω, (c) 12.02 Ω, (d) 16.71 Ω.

8 The $(R_1 + R_2)$ resistance value of 176 m of PVC insulated copper cable having a 2.5 mm^2 phase and protective conductor is
(a) 2.608 Ω, (b) 7.41 Ω, (c) 14.82 Ω, (d) 19.51 Ω.

9 The value of the earth fault loop impedance

Z_s of a circuit fed by 40 m of PVC insulated copper cable having a 2.5 mm^2 phase conductor and1.5 mm^2 protective conductor connected to a supply having an impedance Z_e of 0.5 Ω under fault conditions will be
(a) 1.58 Ω, (b) 9.755 Ω, (c) 20.01 Ω, (d) 780.4 mΩ.

10 The time/current characteristics shown in Figure 3.2 indicate that a fault current of 300 A will cause a 30 A semi-enclosed fuse to BS 3036 to operate in
(a) 0.01 s, (b) 0.1 s, (c) 0.2 s, (d) 2.0 s.

11 The time/current characteristics shown in Figure 3.2 indicate that a fault current of 30 A will cause a 10 A type 2 MCB to BS 3871 to operate in
(a) 0.02 s, (b) 8 s, (c) 33 s, (d) 200 s.

12 'Under fault conditions the protective device nearest to the fault should operate leaving other healthy circuits unaffected'. This is one definition of
(a) fusing factor, (b) effective discrimination, (c) a miniature circuit breaker, (d) a circuit protective conductor.

13 The overcurrent protective device protecting socket outlet circuits and any fixed equipment in bathrooms must operate within
(a) 0.02 s, (b) 0.4 s, (c) 5 s, (d) 45 s.

14 The overcurrent protective device protecting fixed equipment in rooms other than bathrooms must operate within
(a) 0.02 s, (b) 0.4 s, (c) 5 s, (d) 45 s.

15 A 50 mm^2 PVC insulated cable with copper conductors is subjected to a short circuit when connected to a 415 V supply. The impedance of the fault path is 83 mΩ. Calculate the maximum operating time of the protective device to prevent damage to the cable.

16 Explain why the maximum values of earth fault loop impedance Z_s specified by the IEE Regulations and given in Tables 41B1 and 41B2 should not be exceeded.

17 By referring to the tables in Table 41B1, 41B2 and 41D determine the maximum permitted earth fault loop impedance Z_s for the following circuits.
(a) a ring main of 13 A socket outlets protected by a 30 A semi-enclosed fuse to BS 3036, (b) a ring main of 13 A socket outlets protected by a 30 A cartridge fuse to BS 1361, (c) a single socket outlet protected by a 15 A type 1 MCB to BS 3871, (d) a water heating circuit protected by a 15 A semi-enclosed fuse to BS 3036, (e) a lighting circuit protected by a 6 A HBC fuse to BS 88 part 2, (f) a lighting circuit protected by a 5 A semi-enclosed fuse to BS 3036.

18 10 mm^2 cables with PVC insulated copper conductors feed a commercial cooker connected to a 415 V supply. An earth loop impedance test indicates that Z_S has a value of 1.5 Ω. Calculate the minimum size of the protective conductor.

19 It is proposed to protect the commercial cooker circuit described in question 18 with 30 A
(a) semi-enclosed fuses to BS 3036, (b) type 2 MCBs to BS 3871.
Determine the time taken for each protection device to clear an earth fault on this circuit by referring to the characteristics of Figure 3.2.

20 A 2.5 mm^2 PVC insulated and sheathed cable is used to feed a single 13 A socket outlet from a 15 A semi-enclosed fuse in a consumers unit connected to a 240 V supply. Calculate the minimum size of the protective conductor to comply with the Regulations given that the value of Z_S was 0.9 Ω.

CHAPTER 4

Special installations

Wiring circuits and the installation of wiring systems were dealt with in Chapters 7 and 8 of *Basic Electrical Installation Work*. All electrical installations and installed equipment must be safe to use and free from the dangers of an electric shock, but some installations require special consideration because of the inherent dangers of the installed conditions. The danger may arise because of the corrosive or explosive nature of the atmosphere, because the installation must be used in damp or low temperature conditions or because there is a need to provide additional mechanical protection for the electrical system. In this chapter we will consider some of the installations which require special consideration.

Temporary installations

Temporary electrical supplies provided on construction sites can save many man-hours of labour by providing the energy required for fixed and portable tools and lighting which speeds up the completion of a project. However, construction sites are dangerous places and the temporary electrical supply which is installed to assist the construction process must comply with all of the relevant wiring regulations for permanent installations (Regulation 110–01–01). All equipment must be of a robust construction in order to fulfil the on-site electrical requirements whilst being exposed to rough handling, vehicular nudging, the wind, rain and sun. All socket outlets, plugs and couplers must be of the industrial type to BS 4343 and specified by Regulation 604–12–02 as shown in Figure 4.1.

Where an electrician is not permanently on site, MCBs are preferred so that overcurrent protection devices can be safely reset by an unskilled person. The British Standards Code of Practice 1017 'The distribution of electricity on construction and building sites' advises that protection against earth faults may be obtained by first providing a low impedance path, so that overcurrent devices can operate quickly as described in Chapter 3, and secondly by fitting a Residual Current Circuit Breaker in addition to the overcurrent protection device. The 16th Edition of the IEE Regulations considers construction sites very special locations, devoting the whole of Section 604 to their requirements. They have their own set of Tables for disconnection times and maximum earth fault loop impedences (Regulation 604–04–03 and 04). A construction site installation should be tested and inspected in accordance with Part 7 of the Wiring Regulations every three months throughout the construction period.

The source of supply for the temporary installation may be from a petrol or diesel generating set or from the local supply authority. When the local electricity board provides the supply, the incoming cable must be terminated in a waterproof and locked enclosure to prevent unauthorised access and provide metering arrangements.

IEE Regulation 604–02–02 recommend the following voltages for the distribution of electrical supplies to plant and equipment on construction sites:

415 V three phase for supplies to major items of plant having a rating above 3.75 kW such as cranes and lifts. These supplies must be wired in armoured cables.

240 V single phase for supplies to items of equipment which are robustly installed such as floodlighting towers, small hoists and site offices. These supplies must be wired in armoured cable unless run inside the site offices.

110 V single phase for supplies to all portable

hand tools and all portable lighting equipment. The supply is usually provided by a reduced voltage distribution unit which incorporates splashproof sockets fed from a centre-tapped 110 V transformer. This arrangement limits the voltage to earth to 55 V which is recognised as safe in most locations. A 110 V distribution unit is shown in Figure 4.1. Edison screw lamps are used for 110 V lighting supplies so that they are not interchangeable with 240 V site office lamps.

There are occasions when even a 110 V supply from a centre-tapped transformer is too high, for example, supplies to inspection lamps for use inside damp or confined places. In these circumstances a safety extra low voltage SELV supply would be required. This would be provided by a double wound transformer with the secondary isolated from earth as described by Section 411 of the Wiring Regulations.

Industrial plugs have a keyway which prevents a tool from one voltage being connected to the socket outlet of a different voltage. They are also colour coded for easy identification as follows:

415 V – red
240 V – blue
110 V – yellow
50 V – white
25 V – violet

Agricultural and horticultural installations

Especially adverse installation conditions are to be encountered on agricultural and horticultural installations because of the presence of livestock, vermin, dampness, corrosive substances and mechanical damage. The 16th Edition of the IEE Wiring Regulations considers these installations very special locations and has devoted the whole of section 605 to their requirements. They have their own set of Tables for disconnection times and maximum earth fault loop impedances. (605–05–03 and 04). In situations accessible to livestock the electrical equipment should be of Class II construction (double insulated) since it does not rely solely on the installation to provide protection against shock (Regulations 605–11–01).

In buildings intended for livestock, all fixed wiring systems must be inaccessible to the livestock and cables liable to be attacked by vermin must be suitably protected (Regulation 605–12–01 and 02).

Horses and cattle have a very low body resistance which makes them susceptible to an electric shock at voltages lower than 25 V rms, and so where protection is afforded by an overcurrent protective device the values of earth fault loop impedance are reduced as given in Table 605B1 and 605B2. Similarly, where safety extra low voltage, SELV, is used for protection, the upper limit of 50 V a.c. must be reduced appropriately (Regulation 605–02–01).

PVC cables enclosed in heavy duty PVC conduit are suitable for installations in most agricultural buildings. All exposed metalwork must be provided with supplementary equipotential bonding in areas where livestock is kept (Regulation 605–08–02). In many situations water proof sock-

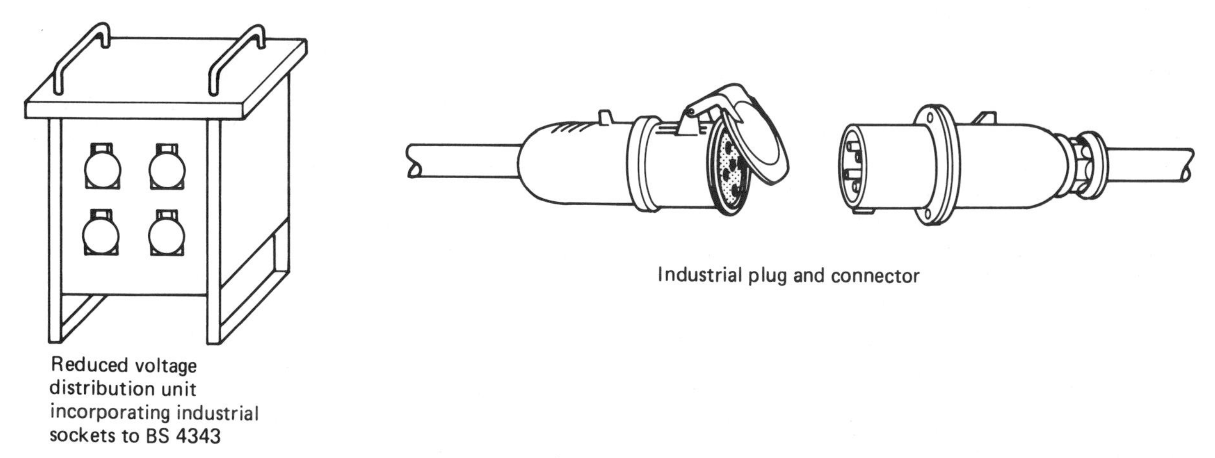

Figure 4.1 110 V distribution unit and cable connector, suitable for construction site electrical supplies

et outlets to BS 196 must be installed together with residual current circuit breaker protection (IEE Regulation 605–03–01).

Cables buried on agricultural or horticultural land should be buried at a depth of not less than 450 mm, or 600 mm where the ground may be cultivated, and the cable must have an armour sheath and be further protected by cable tiles. Overhead cables must be installed so that they are clear of farm machinery or placed at a minimum height of 5.2 m to comply with Regulation 522–08–01 and Table 4B of the On Site Guide.

The sensitivity of farm animals to an electric shock means that they can be contained by an electric fence. An animal touching the fence receives a short pulse of electricity which passes through the animal to the general mass of earth and back to an earth electrode sunk near the controller as shown in Figure 4.2.

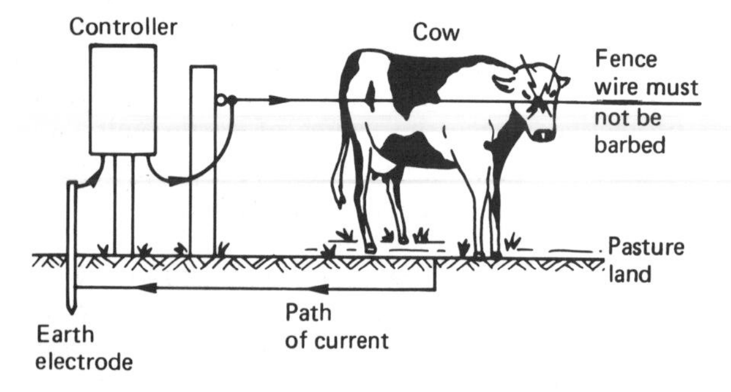

Figure 4.2 Farm animal control by electric fence

The pulses are generated by an RC circuit inside the controller which may be mains or battery operated. There must be no risk to any human coming into contact with the controller, which should be manufactured to BS 2632. Earth electrodes connected to the earth terminal of an electric fence controller must be separate from the earthing system of any other circuit and should be situated outside the resistance area of any electrode used for protective earthing. The electric fence controller and the fence wire must be installed so that they do not come into contact with any power, telephone or radio systems, including poles (Regulation 605–14). Agricultural and horticultural installations should be tested and inspected in accordance with Part 7 of the Wiring Regulations every three years.

Caravans and caravan sites

The electrical installations on caravan sites, and within caravans, must comply in all respects with the wiring regulations for buildings. All the dangers which exist in buildings are present in and around caravans, including the added dangers associated with repeated connection and disconnection of the supply and the flexing of the caravan installation in a moving vehicle. The 16th Edition of the Regulations has devoted Section 608 to the electrical installation in caravans, motor caravans and caravan parks.

Touring caravans must be supplied from an industrial type socket outlet adjacent to the caravan park pitch, similar to that shown in Figure 4.1. Each socket outlet must be supplied, either singly or in groups, with not more than six socket outlets, through a residual current circuit breaker with a rated tripping current of 30 mA. Additionally, every socket outlet must be protected by an overcurrent device (Regulation 608–13–04 and 05). The distance between the caravan and the socket outlet must be not more than 20 m. These requirements are shown in Figure 4.3.

Information notices regarding type and voltage of the supply and maximum permitted load must be displayed at the source of supply. The supply cables must be installed outside the pitch area or be suitably armoured so that they cannot be

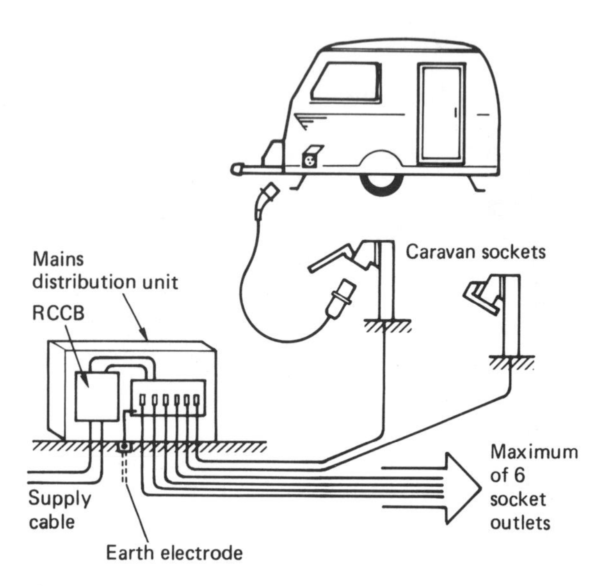

Figure 4.3 Electrical supplies to caravans

damaged by caravan awning pegs (Regulation 608–12–02).

The caravan or motor caravan must be provided with a mains isolating switch and a double pole residual current device (Regulation 608–03–02 and 608–07–04). An adjacent notice detailing how to safely connect and disconnect the supply must also be provided as shown in Regulation 608–07–05. Electrical equipment must not be installed in fuel storage compartments. Caravans flex when being towed, and therefore the installation must be wired in flexible or stranded conductors of at least 1.5 mm cross section. The conductors must be supported on horizontal runs at least every 25 cm and the metalwork of the caravan and chassis must be bonded with 4.0 mm^2 cable.

The wiring of the extra low voltage battery supply must be run in such a way that it does not come into contact with the 240 V wiring system (Regulation 608–06–04)

The caravan should be connected to the pitch socket outlet by means of a flexible cable, not longer than 25 m, and having a minimum CSA of 2.5 mm^2 or as detailed in Table 60A.

Because of the mobile nature of caravans it is recommended that the electrical installation be tested and inspected at periods considered appropriate, between one and three years but not exceeding three years (Regulation 608–07–05).

Flammable and explosive installations

Most flammable liquids only form an explosive mixture between certain concentrated limits. Above and below this level of concentration the mix will not explode. The lowest temperature at which sufficient vapour is given off from a flammable substance to form an explosive gas-air mixture is called the *flashpoint*. A liquid which is safe at normal temperatures will require special consideration if heated to flashpoint. An area in which an explosive gas-air mixture is present is called a *hazardous area*, as defined by BS 5345, and any electrical apparatus or equipment within a hazardous area must be classified as flameproof.

Flameproof electrical equipment is constructed so that it can withstand an internal explosion of the gas for which it is certified, and prevent any spark or flame resulting from that explosion leaking out and igniting the surrounding atmosphere. This is achieved by manufacturing flameproof equipment to a robust standard of construction: All access and connection points have wide machined flanges which damp the flame in its passage across the flange. Flanged surfaces are firmly bolted together with many recessed bolts as shown in Figure 4.4. Wiring systems within a hazardous area must be to flameproof fittings using an appropriate method, such as:

- PVC cables encased in solid drawn heavy gauge screwed steel conduit terminated at approved enclosures having wide flanges and bolted covers.
- Mineral insulated cables terminated into accessories with approved flameproof glands. These have a longer gland thread than normal

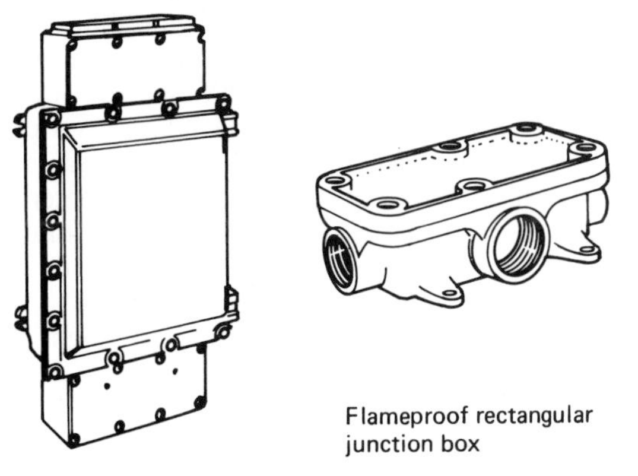

Flameproof distribution board

Flameproof rectangular junction box

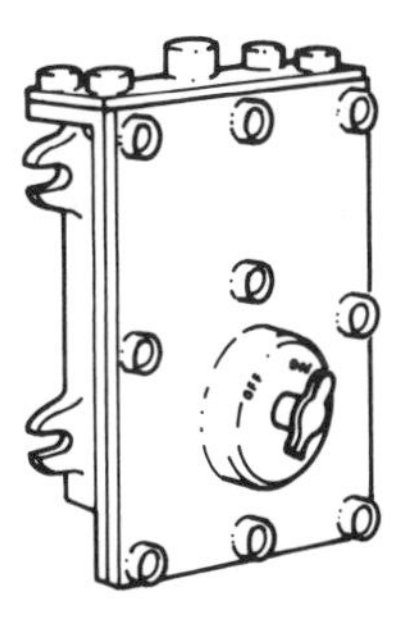

Double-pole switch

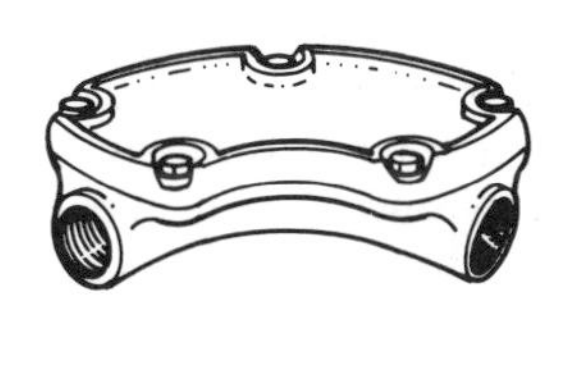

Flameproof inspection bend

Figure 4.4 Flameproof fittings

MICC glands of the type shown in Figure 6.10 of *Basic Electrical Installation Work*. Where the cable is laid underground it must be protected by a PVC sheath and laid at a depth of not less than 500 mm.

- PVC armoured cables terminated into accessories with approved flameproof glands or any other wiring system which is approved by BS 5345. All certified flameproof enclosures will be marked Ex indicating that they are suitable for potentially explosive situations or EEx where equipment is certified to the harmonised European standard.

Flammable and explosive installations are to be found in the petroleum and chemical industries, which are classified as Group 11 industries. Mining is classified as Group 1 and receives special consideration from the Mining Regulations because of the extreme hazards of working underground. Petrol filling pumps must be wired and controlled by flameproof equipment to BS Code of Practice 1003 and meet the requirements of the Petroleum Regulation Act 1928 and 1936 and any local licensing laws concerning the keeping and dispensing of petroleum spirit.

Exercises

1. The temporary electrical installation on a construction site must be inspected and tested:
(a) every three weeks, (b) every month, (c) every three months, (d) at least once each year.
2. Portable hand tools on construction sites should be supplied at:
(a) 50 V, (b) 110 V, (c) 240 V, (d) 415 V.
3. Industrial socket outlet and plugs are colour coded for easy identification, 415 V, 240 V and 110 V plugs are respectively colour coded
(a) red blue and yellow, (b) white blue and green, (c) yellow blue and white, (d) blue yellow and red.
4. Agricultural and horticultural electrical installations must be tested and inspected every:
(a) three months, (b) year, (c) three years,
(d) five years.
5. Mobile caravan electrical installations must be tested and inspected:
(a) before every road journey, (b) at least once each year, (c) every 3 months, (d) at least every 3 years.
6. Caravan site electrical installations must be tested and inspected at least once every
(a) 3 months, (b) year, (c) 3 years, (d) 5 years.
7. A large construction site is provided with a 415 V/240 V supply. Describe with simple sketches how the electrical contractor would supply the following loads from the main supply
(a) a 415 V crane, (b) robustly installed 240 V perimeter lighting, (c) 110 V sockets for portable tools, (d) SELV to special hand lamps, (e) 240 V to site offices.
State the type of cable to be used for each supply.
8. State the precautions to be considered by an electrical contractor asked to wire sockets and lights in a farm building which will be used to accommodate animals.
9. Describe the installation and operation of an electric fence controller.
10. An electrical contractor was asked to design an electrical installation which would be suitable for a 10 van caravan site. Each van was to be supplied with a 240 V socket outlet adjacent to the van and the main electrical supply was to be contained in a central services brick building. Sketch the arrangements and describe the type of cable, socket outlets and control and protection equipment to be used to wire the caravan socket outlets.
11. Describe the type of cable and equipment to be used to wire a petrol pump on a garage forecourt.

CHAPTER 5

a.c. Theory

In Chapter 3 of *Basic Electrical Installation Work* we considered a.c. circuits which were purely resistive, purely inductive or purely capacitive. In practical circuits, at least two and probably all three of these components will be present and it is these circuits which we will now consider.

Opposition to current flow

The opposition to current flow in a purely resistive circuit is known as resistance, measured in ohms and given the symbol R.

$$R = \frac{V_R}{I_R}\ (\Omega).$$

In a purely resistive a.c. circuit the current and voltage are in phase and this can be shown on a phasor diagram illustrated in Figure 5.1(a).

In a purely inductive circuit, that is one which contains a coil or inductor, the opposition to current flow is called inductive reactance and given the symbol X_L.

$$X_L = \frac{V_L}{I_L}\ (\Omega) \qquad \text{or } X_L = 2\ \pi f L\ (\Omega).$$

In a purely inductive circuit the current lags the voltage by 90° as shown in Figure 5.1(b).

In a circuit containing purely capacitance, the opposition to current flow is called capacitive reactance and given the symbol X_C.

$$X_C = \frac{V_C}{I_C}\ (\Omega) \qquad \text{or } X_C = \frac{1}{2\ \pi f C}\ (\Omega).$$

In a purely capacitive circuit the current leads the voltage by 90° as shown in Figure 5.1(c).

When circuits contain two or more separate elements such as RL, RC or RLC, the total opposition to current flow is known as the impedance of the circuit and given the symbol Z.

$$Z = \frac{V_T}{I_T}\ (\Omega) \qquad \text{or } Z = \sqrt{R^2 + X^2}\ (\Omega).$$

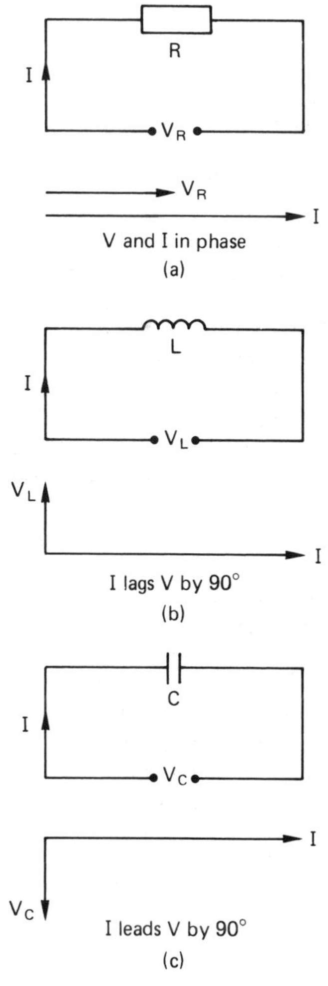

Figure 5.1 Voltage and current characteristics

The phase angle between the total voltage and total current will be neither 0° or 90° but will be determined by the relative values of resistance and reactance in the circuit. In Figure 5.2 the phase angle between applied voltage and current is some angle ϕ.

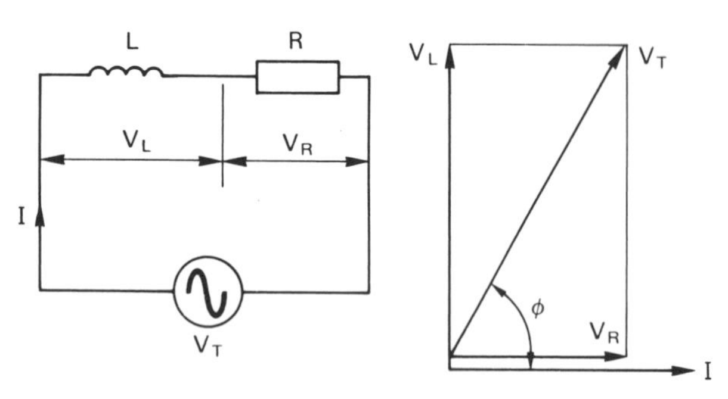

Figure 5.2 A series R-L circuit and phasor diagram

a.c. series circuit

In a circuit containing a resistor and inductor connected in series as shown in Figure 5.2, the current I will flow through the resistor and the inductor causing the voltages V_R to be dropped across the resistor and V_L to be dropped across the inductor. The sum of these voltages will be equal to the total voltage V_T but because this is an a.c. circuit the voltages must be added by phasor addition. The result is shown in Figure 5.2 where V_R is drawn to scale and in phase with the current and V_L is drawn to scale and leading the current by 90°. The phasor addition of these two voltages gives us the magnitude and direction of V_T which leads the current by some angle ϕ.

In a circuit containing a resistor and capacitor connected in series as shown in Figure 5.3, the current I will flow through the resistor and capacitor causing voltage drops V_R and V_C. The voltage V_R will be in phase with the current and V_C will lag the current by 90°. The phasor addition of these voltages is equal to the total voltage V_T which, as can be seen in Figure 5.3, is lagging the current by some angle ϕ.

The impedance triangle

We have now established the general shape of the phasor diagram for a series a.c. circuit. Figures 5.2 and 5.3 show the voltage phasors, but we

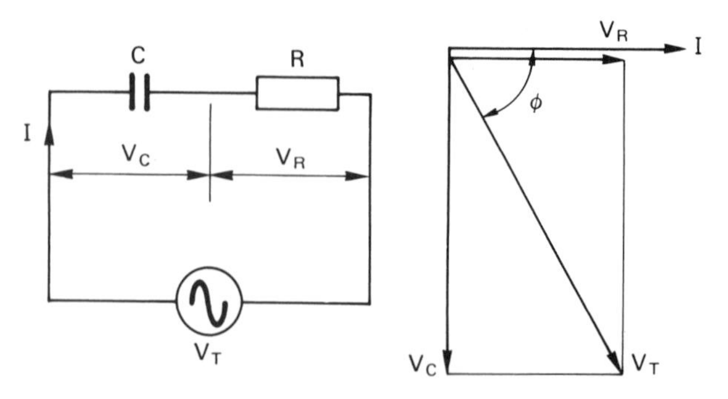

Figure 5.3 A series R-C circuit and phasor diagram

know that $V_R = IR$, $V_L = IX_L$, $V_C = IX_C$ and $V_T = IZ$, and therefore the phasor diagrams (a) and (b) of Figure 5.4 must be equal. From Figure 5.4(b), by the theorem of Pythagoras, we have

$$(IZ)^2 = (IR)^2 + (IX)^2$$

$$I^2Z^2 = I^2R^2 + I^2X^2.$$

If we now divide throughout by I^2 we have

$$Z^2 = R^2 + X^2$$

or $Z = \sqrt{R^2 + X^2}$ (Ω).

The phasor diagram can be simplified to the impedance triangle given in Figure 5.4(c).

Example 1

A coil of 0.15 H is connected in series with a 50 Ω resistor across a 100 V 50 Hz supply. Calculate (a) the reactance of the coil (b) the impedance of the circuit and (c) the current.

For (a), $X_L = 2\pi fL$ (Ω)

$\therefore \quad X_L = 2 \times 3.142 \times 50\text{ Hz} \times 0.15\text{ H} = 47.1\ \Omega.$

For (b), $Z = \sqrt{R^2 + X^2}$ (Ω)

$\therefore \quad Z = \sqrt{(50\ \Omega)^2 + (47.1\ \Omega)^2} = 68.69\ \Omega.$

For (c), $I = V/Z$ (A)

$$\therefore \quad I = \frac{100\text{ V}}{68.69\ \Omega} = 1.46\text{ A}.$$

Example 2

A 60 μF capacitor is connected in series with a 100 Ω resistor across a 240 V 50 Hz supply. Calculate (a) the reactance of the capacitor (b) the impedance of the circuit and (c) the current.

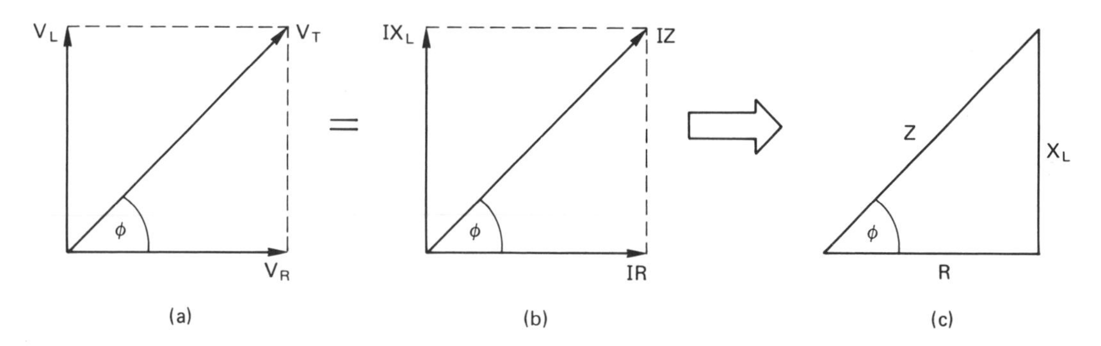

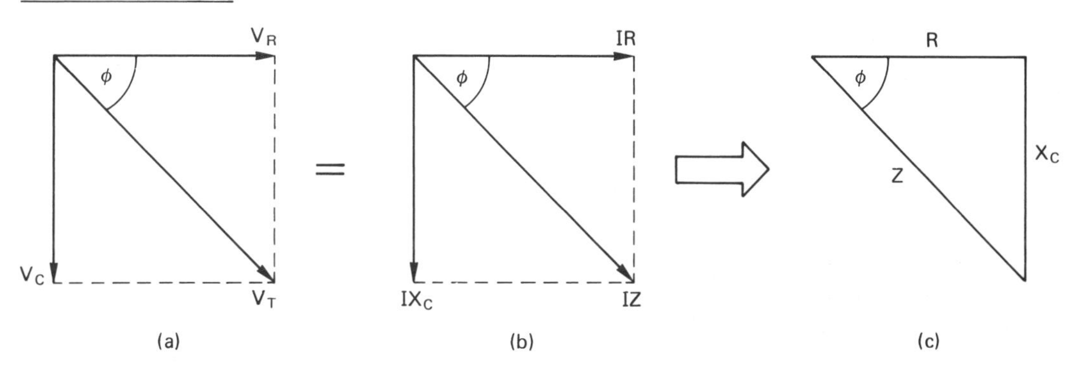

Figure 5.4 Phasor diagram and impedance triangle

For (a), $X_C = \dfrac{1}{2\,\pi f C}\ (\Omega)$

$$\therefore \quad X_C = \frac{1}{2\,\pi \times 50\ \text{Hz} \times 60 \times 10^{-6}\ \text{F}}$$

$$= 53.05\ \Omega.$$

For (b), $Z = \sqrt{R^2 + X^2}\ (\Omega)$

$$\therefore \quad Z = \sqrt{(100\ \Omega)^2 + (53.05\ \Omega)^2}$$

$$= 113.2\ \Omega.$$

For (c), $I = V/Z\ (\text{A})$

$$\therefore \quad I = \frac{240\ \text{V}}{113.2\ \Omega} = 2.12\ \text{A}.$$

Power and power factor

In *Basic Electrical Installation Work* power factor was defined as the cosine of the phase angle between the current and voltage. Power factor may be abbreviated to p.f. If the current lags the voltage as shown in Figure 5.2, we say that the p.f. is lagging, and if the current leads the voltage as shown in Figure 5.3 the p.f. is said to be leading.

$$p.f. = \cos\phi$$

From the trigonometry of the impedance triangle shown in Figure 5.4, p.f. is also equal to:

$$p.f. = \cos\phi = \frac{R}{Z} = \frac{V_R}{V_T}$$

The electrical power in a circuit is the product of the instantaneous value of the voltage and current. Figure 5.5 shows the voltage and current waveform for a pure inductor and pure capacitor. The power waveform is obtained from the product of V and I at every instant in the cycle. It can be seen that the power waveform reverses every quarter cycle, indicating that energy is alternately being fed into and taken out of the inductor and capacitor. When considered over one complete cycle, the

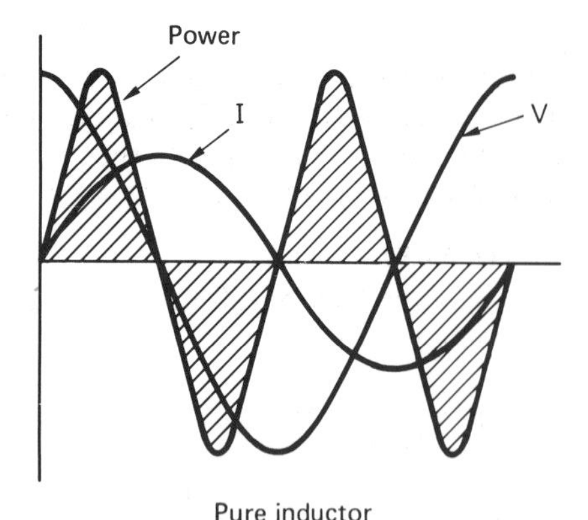

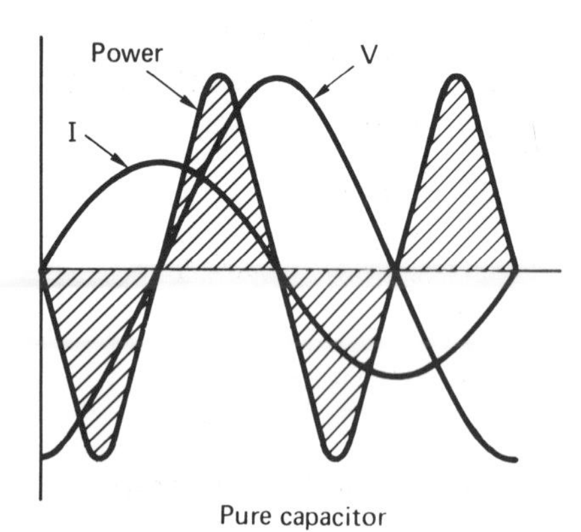

Figure 5.5 Waveform for the a.c. power in purely inductive and purely capacitive circuits

positive and negative portions are equal, showing that the average power consumed by a pure inductor or capacitor is zero. This shows that inductors and capacitors store energy during one part of the voltage cycle and feed it back into the supply later in the cycle. Inductors store energy as a magnetic field and capacitors as an electric field.

In an electric circuit more power is taken from the supply than is fed back into it, since some power is dissipated by the resistance of the circuit and therefore,

$P = I^2 R$ (W).

In any d.c. circuit the power consumed is given by the product of the voltage and current, because in a d.c. circuit voltage and current are in phase. In an a.c. circuit the power consumed is given by the product of the current and that part of the voltage which is in phase with the current. The in phase component of the voltage is given by $V \cos \phi$ and so power can also be given by the equation:

$P = VI \cos \phi$ (W)

Example 1

A coil has a resistance of 30 Ω and a reactance of 40 Ω when connected to a 250 V supply. Calculate (a) the impedance (b) the current (c) the p.f. and (d) the power.

For (a), $Z = \sqrt{R^2 + X^2}\ (\Omega)$

$\therefore\ Z = \sqrt{(30\ \Omega)^2 + (40\ \Omega)^2} = 50\ \Omega.$

For (b), $I = V/Z$ (A)

$\therefore\ I = \dfrac{250\ \text{V}}{50\ \Omega} = 5\ \text{A}.$

For (c), $p.f. = \cos \phi = \dfrac{R}{Z}$

$\therefore\ p.f. = \dfrac{30\ \Omega}{50\ \Omega} = 0.6$ lagging.

For (d), $P = VI \cos \phi$ (W)

$\therefore\ P = 250\ \text{V} \times 5\ \text{A} \times 0.6 = 750\ \text{W}.$

Example 2

A capacitor of reactance 12 Ω is connected in series with a 9 Ω resistor across a 150 V supply. Calculate (a) the impedance of the circuit (b) the current (c) the p.f. and (d) the power.

For (a), $Z = \sqrt{R^2 + X^2}\ (\Omega)$

$\therefore\ Z = \sqrt{(9\ \Omega)^2 + (12\ \Omega)^2} = 15\ \Omega.$

For (b), $I = V/Z$ (A)

$\therefore\ I = \dfrac{150\ \text{V}}{15\ \Omega} = 10\ \text{A}.$

For (c), $p.f. = \cos \phi = \dfrac{R}{Z}$

$\therefore\ p.f. = \dfrac{9\ \Omega}{15\ \Omega} = 0.6$ leading.

For (d), $P = VI \cos \phi$ (W)

$\therefore\ P = 150\ \text{V} \times 10\ \text{A} \times 0.6 = 900\ \text{W}.$

Resistance inductance and capacitance in series

The circuit diagram and phasor diagram are shown in Figure 5.6. The voltages across the components are represented by V_R, V_L and V_C which have the directions shown. Since V_L leads I by 90° and V_C lags by 90° the phasors are in opposition and the combined result is given by $V_L - V_C$ as shown.

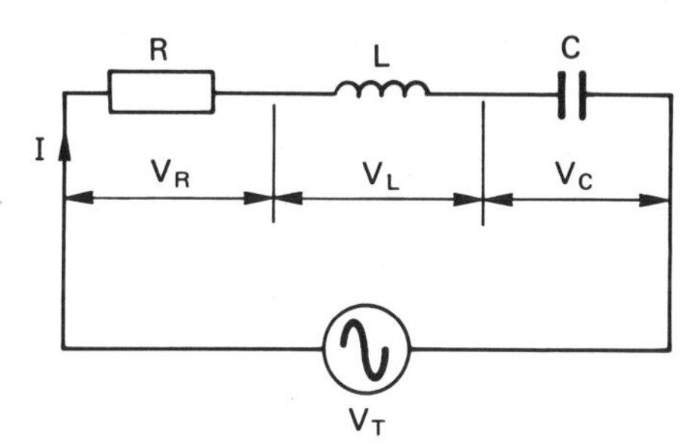

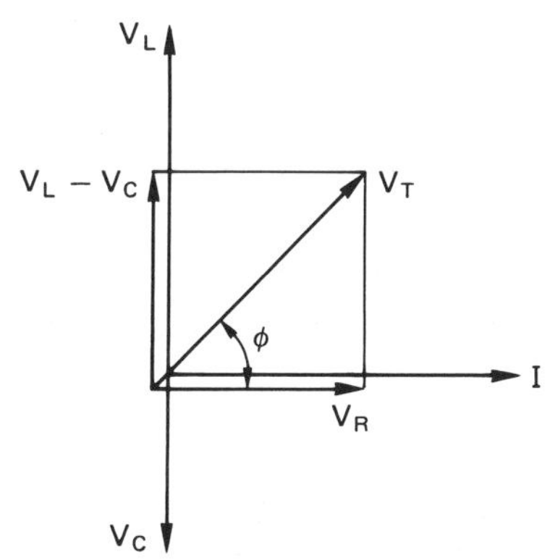

Figure 5.6 R-L-C series circuit and phasor diagram

Applying the theorem of Pythagoras to the phasor diagram of Figure 5.6 we have

$V_T{}^2 = V_R{}^2 + (V_L - V_C)^2.$

Since $V_T = IZ$, $V_R = IR$, $V_L = IX_L$ and $V_C = IX_C$ this equation may also be expressed thus:

$(IZ)^2 = (IR)^2 + (IX_L - IX_C)^2.$

Cancelling out the common factors we have

$$Z^2 = R^2 + (X_L - X_C)^2$$

or $Z = \sqrt{R^2 + (X_L - X_C)^2}$ (Ω).

Note The derivation of this equation is not required by craft students but the equation should be remembered and applied in appropriate cases.

Example 1

A coil of resistance 5 Ω and inductance 10 mH is connected in series with a 75 μF capacitor across a 200 V 100 Hz supply. Calculate (a) the impedance of the circuit (b) the current and (c) the p.f..

For (a), $X_L = 2\pi fL$ (Ω)

$\therefore\ X_L = 2 \times 3.142 \times 100\text{ Hz} \times 10 \times 10^{-3}\text{ H}$

$X_L = 6.28\ \Omega$

$X_C = \dfrac{1}{2\pi fC}$ (Ω)

$\therefore\ X_C = \dfrac{1}{2 \times 3.142 \times 100\text{ Hz} \times 75 \times 10^{-6}\text{ F}}$

$X_C = 21.22\ \Omega$

$Z = \sqrt{R^2 + (X_L - X_C)^2}$ (Ω)

$\therefore\ Z = \sqrt{(5\ \Omega)^2 + (6.28\ \Omega - 21.22\ \Omega)^2}$

$Z = 15.75\ \Omega.$

For (b), $I = V/Z$ (A)

$\therefore\ I = \dfrac{200\text{ V}}{15.75\ \Omega} = 12.69\text{ A}.$

For (c), $p.f. = \cos\phi = \dfrac{R}{Z}$

$\therefore\ p.f. = \dfrac{5}{15.7} = 0.317$ Leading.

Example 2

A 200 μF capacitor is connected in series with a coil of resistance 10 Ω and inductance 100 mH to a 240 V 50 Hz supply. Calculate (a) the impedance (b) the current and (c) the voltage dropped across each component.

For (a), $X_L = 2\pi fL$ (Ω)

$\therefore\ X_L = 2 \times 3.142 \times 50\text{ Hz} \times 100 \times 10^{-3}\text{ H}$

$X_L = 31.42\ \Omega$

$X_C = \dfrac{1}{2\pi fC}$ (Ω)

$\therefore\ X_C = \dfrac{1}{2 \times 3.142 \times 50\text{ Hz} \times 200 \times 10^{-6}}$

$X_C = 15.9\ \Omega$

$Z = \sqrt{R^2 + (X_L - X_C)^2}$ (Ω)

$\therefore\ Z = \sqrt{(10\ \Omega)^2 + (31.42\ \Omega - 15.9\ \Omega)^2}$

$Z = 18.46\ \Omega.$

For (b), $I = V/Z$ (A)

$$\therefore \quad I = \frac{240\ \text{V}}{18.46\ \text{A}} = 13\ \text{A}.$$

For (c), $V_R = I \times R$ (V)

$$\therefore \quad V_R = 13\ \text{A} \times 10\ \Omega = 130\ \text{V}$$

$$V_L = I \times X_L\ \text{(V)}$$

$$\therefore \quad V_L = 13\ \text{A} \times 31.42\ \Omega = 408.46\ \text{V}$$

$$V_C = I \times X_C\ \text{(V)}$$

$$\therefore \quad V_C = 13\ \text{A} \times 15.9\ \Omega = 206.7\ \text{V}$$

The phasor diagram of this circuit would be similar to that shown in Figure 5.6.

Series resonance

At resonance the circuit responds sympathetically. Therefore, the condition of resonance is used extensively in electronic and communication circuits for frequency selection and tuning. The current and reactive components of the circuit are at a maximum and so resonance is usually avoided in power applications to prevent cables being overloaded and cable insulation being broken down.

A circuit can be tuned to resonance by either varying the capacitance of the circuit or by adjusting the supply frequency. At low frequencies the circuit is mainly capacitive and at high frequencies the inductive effect predominates. At some intermediate frequency a point exists where the capacitive effect exactly cancels the inductive effect. This is the point of resonance and occurs when

$$V_L = V_C.$$

$$\therefore \quad IX_L = IX_C.$$

If we cancel the common factor we have

$$X_L = X_C$$

$$\therefore \quad 2\pi fL = \frac{1}{2\pi fC}.$$

Collecting terms

$$f^2 = \frac{1}{4\pi^2 LC}.$$

Taking square roots,

$$\text{Resonant frequency} = f_0 = \frac{1}{2\pi}\sqrt{\frac{1}{LC}}\ \text{(Hz)}.$$

Note The resonant frequency is given the symbol f_0. The derivation of the formulae is not required by craft students.

At resonance the circuit is purely resistive. $Z = R$, the phase angle is zero and therefore the supply voltage and current must be in phase. These effects are shown in Figure 5.7.

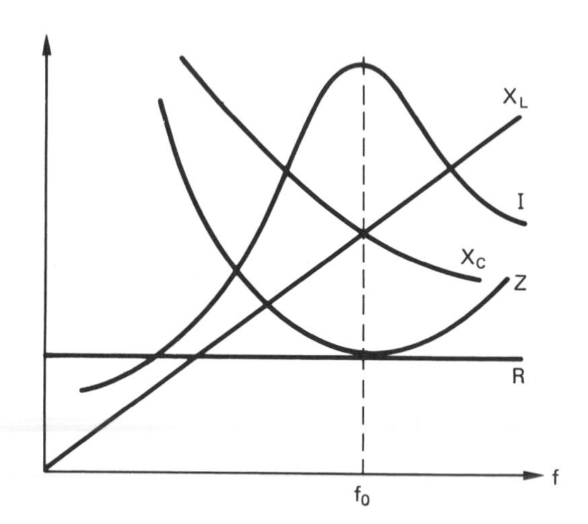

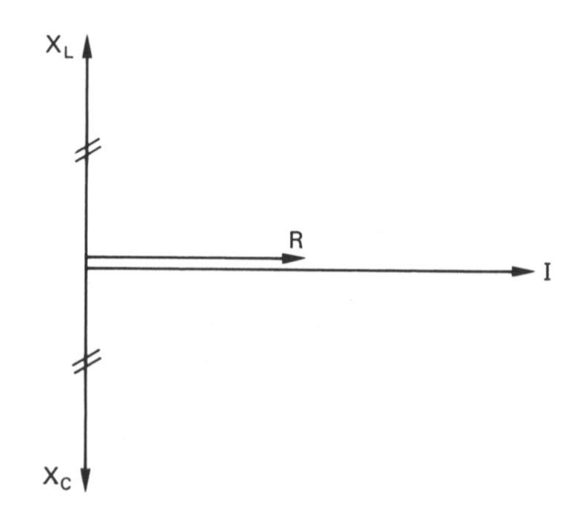

Figure 5.7 Series resonance conditions in an R-L-C circuit

Example 1

A capacitor is connected in series with a coil of resistance 50 Ω and inductance 168.8 mH across a 50 Hz supply. Calculate the value of the capacitor to produce resonance in this circuit.

$$X_L = 2\pi fL \ (\Omega)$$
$$\therefore \quad X_L = 2 \times 3.142 \times 50 \text{ Hz} \times 168.8 \times 10^{-3} \text{ H}.$$
$$X_L = 53.03 \ \Omega.$$

At resonance $X_L = X_C$ therefore $X_C = 53.03 \ \Omega$

$$X_C = \frac{1}{2\pi fC} \ (\Omega).$$

Transposing for $C = \dfrac{1}{2\pi fX_C}$ (F)

$$\therefore \quad C = \frac{1}{2 \times 3.142 \times 50 \text{ Hz} \times 53.03 \ \Omega}$$

$$C = 60 \ \mu\text{F}.$$

Example 2
Calculate the resonant frequency of a circuit consisting of a 25.33 mH inductor connected in series with a 100 μF capacitor

Resonant frequency $= f_0 = \dfrac{1}{2\pi}\sqrt{\dfrac{1}{LC}}$ (Hz).

$$\therefore \quad f_0 = \frac{1}{2\pi}\sqrt{\frac{1}{25.33 \times 10^{-3} \text{ H} \times 100 \times 10^{-6} \text{ F}}}$$

$$f_0 = 100 \text{ Hz}.$$

Parallel circuits

In practice, most electrical installations consist of a number of circuits connected in parallel to form a network. The branches of the parallel network may consist of one component or two or more components connected in series. You should now have an appreciation of series circuits and we will now consider two branch parallel circuits. In a parallel circuit the supply voltage is applied to each of the network branches. Voltage is used as the reference when drawing phasor diagrams and the currents are added by phasor addition.

In a parallel circuit containing a pure resistor and inductor as shown in Figure 5.8, the current flowing through the resistive branch will be in phase with the voltage and the current flowing in the inductive branch will be 90° lagging the voltage. The phasor addition of these currents will give the total current drawn from the supply and its phase angle as shown in the phasor diagram of Figure 5.8.

In a parallel circuit containing a pure resistor

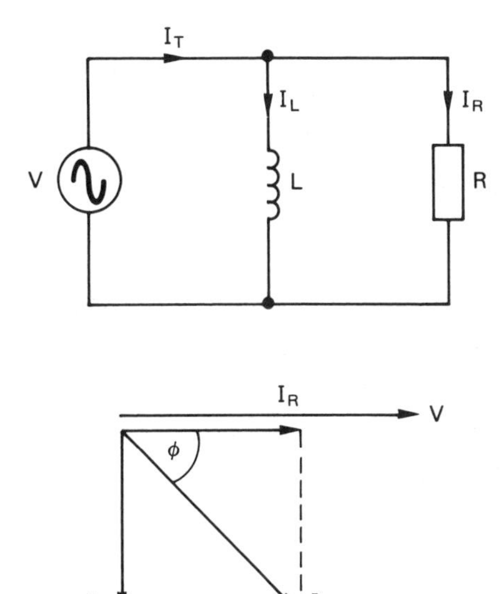

Figure 5.8 A parallel R-L circuit and phasor diagram

and capacitor connected in parallel as shown in Figure 5.9, the current flowing through the resistive branch will be in phase with the voltage and the current in the capacitive branch will lead the voltage by 90°. The phasor addition of these currents will give the total current and its phase angle as shown in Figure 5.9.

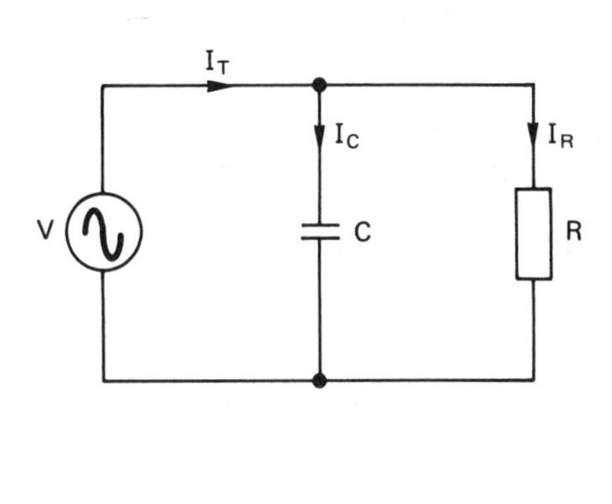

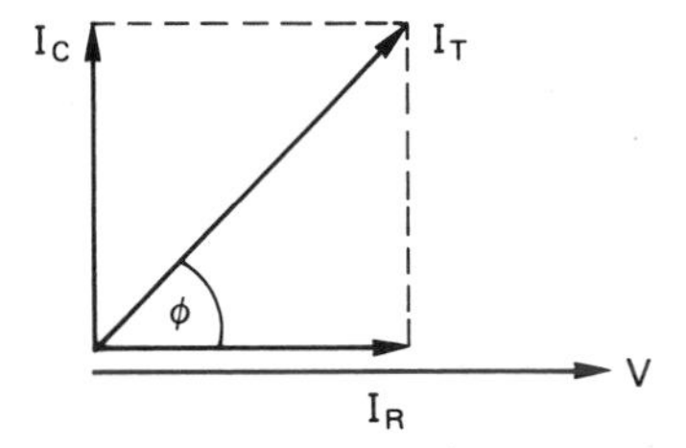

Figure 5.9 A parallel R-C circuit and phasor diagram

Example 1
A pure inductor of 100 mH is connected in parallel with a 30 Ω resistor to a 240 V 50 Hz supply. Calculate the branch currents and the supply current.

$$I_R = \frac{V}{R}\ (A)$$

$$\therefore \quad I_R = \frac{240\ V}{30\ \Omega} = 8\ A$$

$$X_L = 2\pi fL\ (\Omega)$$

$$\therefore \quad X_L = 2 \times 3.142 \times 50\ Hz \times 100 \times 10^{-3}\ H$$

$$X_L = 31.42\ \Omega$$

$$I_L = \frac{V}{X_L}\ (A)$$

$$\therefore \quad I_L = \frac{240\ V}{31.42\ \Omega} = 7.64\ A.$$

From the trigonometry of the phasor diagram in Figure 5.8 the total current is given by

$$I_T = \sqrt{I_R{}^2 + I_L{}^2}\ (A)$$

$$\therefore \quad I_T = \sqrt{(8A)^2 + (7.64A)^2}$$

$$I_T = 11.06\ A.$$

Example 2
A pure capacitor of 60 μF is connected in parallel with a 40 Ω resistor across a 240 V 50 Hz supply. Calculate the branch currents and the supply currents.

$$I_R = \frac{V}{R}\ (A)$$

$$\therefore \quad I_R = \frac{240\ V}{40\ \Omega} = 6\ A$$

$$X_C = \frac{1}{2\pi fC}\ (\Omega)$$

$$\therefore \quad X_C = \frac{1}{2 \times 3.142 \times 50\ Hz \times 60 \times 10^{-6}\ F}$$

$$X_C = 53.05\ \Omega$$

$$I_C = \frac{V}{X_C}\ (A)$$

$$\therefore \quad I_C = \frac{240\ V}{53.05\ \Omega} = 4.52\ A.$$

From the trigonometry of the phasor diagram in Figure 5.9 the total current is given by

$$I_T = \sqrt{I_c^2 + I_R^2}\ (A)$$

$$I_T = \sqrt{(6A)^2 + (4.52\ A)^2}$$

$$I_T = 7.5\ A.$$

In considering these two examples we have assumed the capacitor and inductor to be pure. In practice the inductor will contain some resistance and the network may, therefore, be considered as a series RL branch connected in parallel with a capacitor as shown in Figure 5.10.

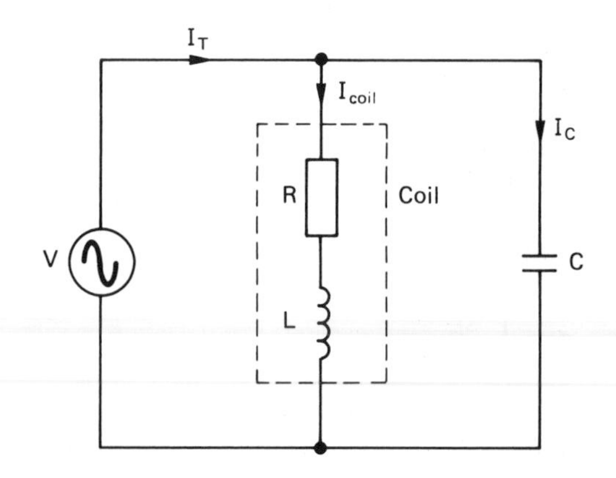

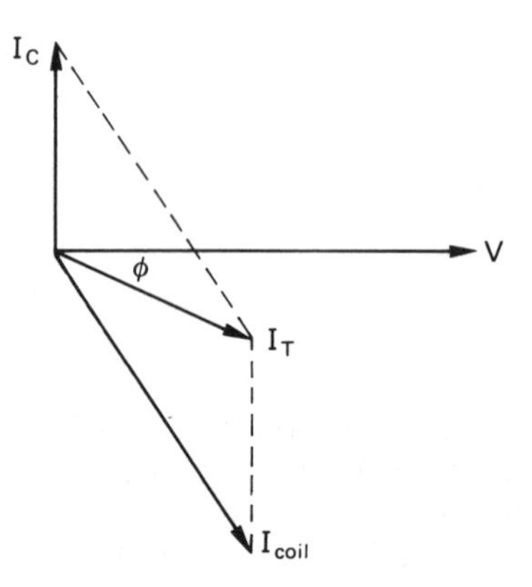

Figure 5.10 A parallel circuit and phasor diagram

Example 3
A coil having a resistance of 50 Ω and inductance 318 mH is connected in parallel with 20 μF capacitor across a 240 V 50 Hz supply. Calculate the branch currents and the supply current. The circuit diagram for this network is shown in Figure 5.10.

$$X_C = \frac{1}{2\pi fC} \ (\Omega)$$

$$\therefore \quad X_C = \frac{1}{2 \times 3.142 \times 50 \text{ Hz} \times 20 \times 10^{-6} \text{ F}}$$

$$X_C = 159.2 \ \Omega$$

$$I_C = \frac{V}{X_c} \ (\text{A})$$

$$\therefore \quad I_C = \frac{240 \text{ V}}{159.2 \ \Omega} = 1.5 \text{ A}$$

$$X_L = 2\pi fL \ (\Omega)$$

$$\therefore \quad X_L = 2 \times 3.142 \times 50 \text{ Hz} \times 318 \times 10^{-3} \text{ H}.$$

$$X_L = 100 \ \Omega$$

$$Z_{coil} = \sqrt{R^2 + X_L{}^2} \ (\Omega)$$

$$\therefore \quad Z_{coil} = \sqrt{(50 \ \Omega)^2 + 100 \ \Omega)^2}$$

$$Z_{coil} = 111.8 \ \Omega$$

$$I_{coil} = \frac{V}{Z} \ (\text{A})$$

$$\therefore \quad I_{coil} = \frac{240 \text{ V}}{111.8 \ \Omega} = 2.15 \text{ A}.$$

The capacitor current will lead the supply voltage by 90°. The coil current will lag the voltage by some angle given by:

$$\phi_{coil} = \cos^{-1} \frac{R}{Z}$$

$$\therefore \quad \phi_{coil} = \cos^{-1} \frac{50 \ \Omega}{111.8 \ \Omega} = 63.4^\circ$$

The coil and capacitor currents can now be drawn to scale as shown in Figure 5.11. The total current is the phasor addition of these currents and is found to be 1.08A at 23° lagging from the phasor diagram.

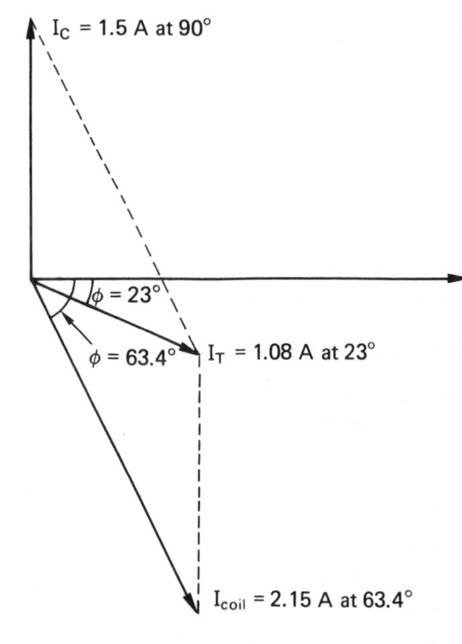

Figure 5.11 Scale phasor diagram for Example 3

Exercises

1 An a.c. series circuit has an inductive reactance of 4 Ω and a resistance of 3 Ω. The impedance of this circuit will be:
(a) 5 Ω, (b) 7 Ω, (c) 12 Ω, (d) 25 Ω.

2 An a.c. series circuit has a capacitive reactance of 12 Ω and a resistance of 9 Ω. The impedance of this circuit will be:
(a) 3 Ω, (b) 15 Ω, (c) 20 Ω, (d) 108 Ω.

3 A circuit whose resistance is 3 Ω, reactance 5 Ω and impedance 5.83 Ω will have a p.f. of
(a) 0.515, (b) 0.600, (c) 0.858, (d) 1.666.

4 A circuit whose resistance is 5 Ω, capacitance reactance 12 Ω and inductive reactance 20 Ω will have an impedance of:
(a) 9.434 Ω, (b) 21.1891, (c) 23.853 Ω, (d) 32.388 Ω.

5 The inductive reactance of a 100 mH coil when connected to 50 Hz will be:
(a) 0.5 Ω, (b) 0.0318 Ω, (c) 5.0 Ω, (d) 31.416 Ω.

6 The capacitive reactance of a 100 μF capacitor connected to a 50 Hz supply will be:
(a) 0.5 Ω, (b) 5.0 mΩ, (c) 31.83 Ω, (d) 31415.93 Ω.

7 In a series resonant circuit the:
(a) current and impedance are equal,
(b) current is at a minimum and the impedance a maximum,

(c) current is at a maximum and the impedance a minimum, (d) current and impedance are at a maximum.

8 In a series resonant circuit the:
(a) capacitive and inductive reactances are equal, (b) capacitive and inductive reactances are at a minimum value, (c) capacitive and inductive reactances are at a maximum value, (d) capacitive and inductive reactances are equal to the resistance of the circuit.

9 A circuit containing a 100 μF capacitor in series with a 100 mH inductor will resonate at a frequency of:
(a) 3.142 Hz, (b) 50.33 Hz, (c) 316.23 Hz, (d) 10 kHz.

10 A series circuit consisting of a 25.33 mH inductor and 100 μF capacitor will have a resonant frequency of:
(a) 2.53 Hz, (b) 79.58 Hz, (c) 90 Hz, (d) 100 Hz.

11 A 10 Ω resistor is connected in series with an inductor of reactance 15 Ω across a 240 V a.c. supply. Calculate
(a) the impedance, (b) the current, (c) the voltage across each component, (d) the power factor, (e) draw the phasor diagram to scale.

12 A 9 Ω resistor is connected in series with a capacitor of 265.25 μF across a 240 V 50 Hz supply. Calculate
(a) the impedance, (b) the current, (c) the voltage across each component, (d) the power factor, (e) draw to scale the phasor diagram.

13 A 15 Ω resistor is connected in series with a 60 μF capacitor and a 0.1 H inductor across a 240 V 50 Hz supply. Draw the circuit diagram and calculate
(a) the impedance, (b) the current, (c) the voltage across each component, (d) the power factor, (e) draw to scale the phasor diagram.

14 Derive from first principles an expression for the resonant frequency of a series circuit.

15 Sketch a graph to show the variations of R, X_L, X_c, Z and I with frequency for a series a.c. circuit. Indicate the point at which resonance occurs on the graph using frequency as a base.

16 A pure inductor of 100 mH is connected in parallel with a 15 Ω resistor across a 240 V 50 Hz supply. Calculate the current in each branch and the total current and power factor. Sketch the phasor diagram.

17 A 60 μF capacitor is connected in parallel with a 20 Ω resistor across a 240 V 50 Hz supply. Calculate the current in each branch and the total current and power factor. Sketch the phasor diagram.

CHAPTER 6

Mechanics

Basic mechanisms and machines were considered briefly in Chapter 4 of *Basic Electrical Installation Work*, but here we will consider some fundamental mechanical principles and calculations.

Fundamental principles

Mass

This is a measure of the amount of material in a substance such as metal, plastic, wood, brick or tissue which is collectively known as a body. The SI unit of mass is the kilogram abbreviated to kg.

Speed

The feeling of speed is something with which we are all familiar. If we travel in a motor vehicle we know that an increase in speed would, excluding accidents, allow us to arrive at our destination more quickly. Therefore, speed is concerned with distance travelled and time taken. Suppose we were to travel a distance of 30 miles in one hour, our speed would be an average of 30 miles per hour:

$$\text{Speed} = \frac{\text{Distance}}{\text{Time}}\,\frac{(\text{m})}{(\text{s})}$$

Velocity

In everyday conversation we often use the word velocity to mean the same as speed, and indeed the units are the same. However, for scientific purposes this is not acceptable since velocity is also concerned with direction. Velocity is speed in a given direction. For example, the speed of an aircraft might be 200 miles per hour, but its velocity would be 200 miles per hour in, say, a westerly direction. Speed is a scalar quantity where velocity is a vector quantity.

$$\text{Velocity} = \frac{\text{Distance}}{\text{Time}}\,\frac{(\text{m})}{(\text{s})}$$

Acceleration

When an aircraft takes off, it starts from rest and increases its velocity until it can fly. This change in velocity is called its acceleration. By definition acceleration is the rate of change in velocity with time.

$$\text{Acceleration} = \frac{\text{Velocity}}{\text{Time}}\ (\text{m/s}^2)$$

The SI unit for acceleration is the metre per second or m/s^2.

Example
If an aircraft accelerates from a velocity of 15 m/s to 35 m/s in 4 s calculate its average acceleration:

$$\text{Average velocity} = 35\ \text{m/s} - 15\ \text{M/s} = 20\ \text{m/s}$$

$$\text{Average acceleration} = \frac{\text{Velocity}}{\text{Time}} = \frac{20}{4} = 5\ \text{m/s}^2.$$

Thus, the average acceleration is 5 metres per second, every second.

Force

The presence of a force can only be detected by its effect on a body. A force may cause a stationary object to move or bring a moving body to rest. For example, a number of people pushing a broken down motor car exert a force which propels it forward, but applying the motor car brakes applies a force on the brake drums which slows down or stops the vehicle. Gravitational force causes objects to fall to the ground. The apple fell from the tree on to Isaac Newton's head as a result of gravitational force. The standard rate of acceleration due to gravity is accepted as 9.81 m/s^2. Therefore, an apple weighing 1 kg will exert a force of 9.81 N since

$$\text{Force} = \text{Mass} \times \text{Acceleration (N)}.$$

The SI unit of force is the newton, symbol N, to commemorate the great English scientist Sir Isaac Newton (1642–1727).

Example
A 50 kg bag of cement falls from a forklift truck whilst being lifted to a storage shelf. Determine the force with which the bag will strike the ground:

$$\text{Force} = \text{Mass} \times \text{Acceleration (N)}$$
$$\text{Force} = 50 \text{ kg} \times 9.81 \text{ m/s}^2 = 490.5 \text{ N}.$$

Pressure or stress

To move a broken motor car I might exert a force on the back of the car to propel it forward. My hands would apply a pressure on the body panel at the point of contact with the car. Pressure or stress is a measure of the force per unit area.

$$\text{Pressure or stress} = \frac{\text{Force}}{\text{Area}} \ (\text{N/m}^2)$$

Example 1
A young woman of mass 60 kg puts all her weight on to the heel of one shoe which has an area of 1 cm^2. Calculate the pressure exerted by the shoe on the floor. (Assuming the acceleration due to gravity to be 9.81 m/s^2).

$$\text{Pressure} = \frac{\text{Force}}{\text{Area}} \ (\text{N/m}^2)$$

$$\text{Pressure} = \frac{60 \text{ kg} \times 9.81 \text{ m/s}^2}{1 \times 10^{-4} \text{ m}^2} = 5886 \text{ kN/m}^2.$$

Example 2
A small circus elephant of mass 1 tonne (1000 kg) puts all its weight on to one foot which has a surface area of 400 cm^2. Calculate the pressure exerted by the elephant's foot on the floor, assuming the acceleration due to gravity to be 9.81 m/s^2.

$$\text{Pressure} = \frac{\text{Force}}{\text{Area}} \ (\text{N/m}^2)$$

$$\text{Pressure} = \frac{1000 \text{ kg} \times 9.81 \text{ m/s}^2}{400 \times 10^{-4} \text{ m}} = 245.3 \text{ kN/m}^2.$$

These two examples show that the young woman exerts 24 times more pressure on the ground than the elephant. This is because her mass exerts a force over a much smaller area than the elephant's foot, and is the reason why many wooden dance floors are damaged by high heeled shoes.

Work done

Suppose a broken motor car was to be pushed along a road, work would be done on the car by applying the force necessary to move it along the road. Heavy breathing and perspiration would be evidence of the work done.

$$\text{Work done} = \text{Force} \times \text{Distance moved in the direction of the force (J)}$$

The SI unit of work done is the Newton metre or Joule. The Joule is the preferred unit and it commemorates an English physicist, James Prescot Joule (1818–1889).

Example
A building hoist lifts ten 50 kg bags of cement through a vertical distance of 30 m to the top of a high rise building. Calculate the work done by the hoist, assuming the acceleration due to gravity to be 9.81 m/s^2.

$$\begin{aligned}
&\text{Work done} = \text{Force} \times \text{Distance moved (J)} \\
\text{but} \quad &\text{Force} = \text{Mass} \times \text{Acceleration (N)} \\
\therefore \quad &\text{Work done} = \text{Mass} \times \text{Acceleration} \times \text{Distance moved (J)} \\
&\text{Work done} = 10 \times 50 \text{ kg} \times 9.81 \text{ m/s}^2 \times 30 \text{ m} \\
&\text{Work done} = 147.15 \text{ kJ}.
\end{aligned}$$

Power

If a motor car can cover the distance between two points more quickly than another car, we say that the faster car is more powerful. It can do a given amount of work more quickly. By definition, power is the rate of doing work.

$$\text{Power} = \frac{\text{Work done}}{\text{Time taken}} \text{ (W)}.$$

The SI unit of power, both electrical and mechanical, is the watt symbol W. This commemorates the name of James Watt (1736–1819) the inventor of the steam engine.

Example 1
A building hoist lifts ten 50 kg bags of cement to the top of a 30 m high building. Calculate the rating (power) of the motor to perform this task in 60 seconds if the acceleration due to gravity is taken as 9.81 m/s^2.

$$\text{Power} = \frac{\text{Work done}}{\text{Time taken}} \text{ (W)},$$

but Work done = Force × Distance moved (J)
and Force = Mass × Acceleration (N).

By substitution,
Power

$$= \frac{\text{Mass} \times \text{Acceleration} \times \text{Distance moved}}{\text{Time taken}} \text{ (W)}$$

$$\text{Power} = \frac{10 \times 50 \text{ kg} \times 9.81 \text{ m/s}^2 \times 30 \text{ m}}{60 \text{ s}}$$

Power = 2452.5 W.

The rating of the building hoist motor will be 2.45 kW.

Example 2
A hydro-electric power station pump motor working continuously during a 7 hour period raises 856 tonne of water through a vertical distance of 60 m. Determine the rating (power) of the motor, assuming the acceleration due to gravity is 9.81 m/s^2.

From Example 1,

Power

$$= \frac{\text{Mass} \times \text{Acceleration} \times \text{Distance moved}}{\text{Taken}} \text{ (W)}$$

Power

$$= \frac{856 \times 1000 \text{ kg} \times 9.81 \text{ m/s}^2 \times 60 \text{ m}}{7 \times 60 \times 60 \text{ s}}$$

Power = 20 000 W.

The rating of the pump motor is 20 kW.

Example 3
An electric hoist motor raises a load of 500 kg at a velocity of 2 m/s. Calculate the rating (power) of the motor if the acceleration due to gravity is 9.81 m/s^2.

Power

$$= \frac{\text{Mass} \times \text{Acceleration} \times \text{Distance moved}}{\text{Time taken}} \text{ (W)}$$

$$\text{but Velocity} = \frac{\text{Distance}}{\text{Time}} \text{ (m/s)}.$$

$\therefore$ Power = Mass × Acceleration × Velocity
Power = 500 kg × 9.81 m/s^2 × 2 m/s
Power = 9810 W.

The rating of the hoist motor is 9.81 kW.

Energy

Have you ever observed how some people seem to have more energy than others? Some footballers appear to be able to run around the sports field all day and remain apparently fresh, whilst others are exhausted after a short dash. The player with the most energy is not only a powerful player but can persist in his efforts for a longer period of time.

Energy = Power × time (J).

Anything which is capable of doing work is said to possess energy and so the SI unit of energy is the same as for work; the joule, symbol J.

Energy in the scientific meaning can take many forms, electrical, mechanical, thermal, chemical and atomic, but in mechanics, energy is divided into two types; potential and kinetic energy.

Potential energy is the energy a body possesses because of its position or state of compression. A body raised some distance above the ground possesses potential energy since its mass can do work as it returns to the ground. A wound up spring possesses potential energy because a clock spring does work against the forces opposing the motion of the clock mechanism.

Kinetic energy is the energy a body possesses because of its motion. The water which flows from the dam of a hydro-electric power station possesses kinetic energy since the rapidly flowing water does work by turning a turbine and generating electricity. A hammer head possesses kinetic energy since it can do work by striking a nail and driving the nail into a piece of wood.

Conservation of energy

As you will be aware, the efficiency of a machine is never 100%. In converting energy from one form to another the useful output is always less than the input. This does not mean that energy is destroyed, only that some of the useful energy is lost in the conversion. In the case of an electrical to mechanical conversion, some of the input energy is changed to heat in overcoming friction.

In all energy conversions some of the useful energy is lost but it is not destroyed. It is merely converted to a form which, in most cases, is undesirable or irretrievable. The Conservation of Energy Theory states that '*energy can neither be created nor destroyed, but can be transferred from one form to another*'.

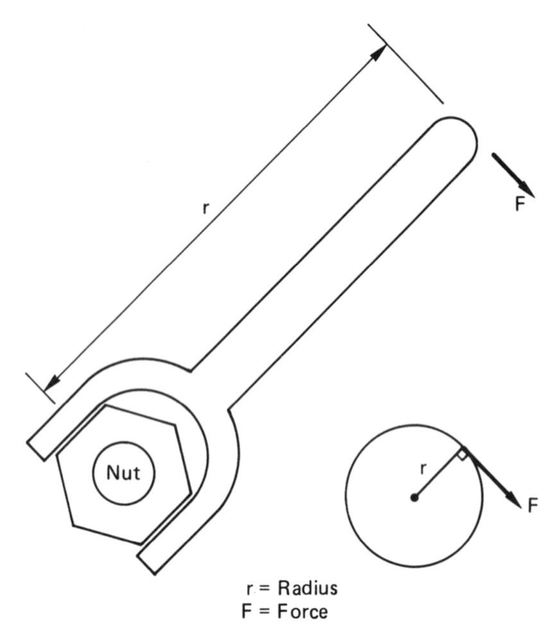

Figure 6.1 Application of a force

Torque

The turning moment of a force applied at a radius is called the torque and is equal to the product of the force applied and the radius.

Torque = Force × Radius (Nm)

If a force is applied to a body at a radius as shown in Figure 6.1, that body will tend to turn. For example, a torque is applied to tighten a nut, turn the starting handle of a car or cause a pulley to rotate.

Example
A pulley of 26 cm diameter is driven by a 'V' belt from an electric motor. If the force tending to turn the pulley is 30 N, calculate the torque on the pulley.

Torque = Force × Radius (Nm)

$$\text{Torque} = 30\ \text{N} \times \frac{13}{100}\ \text{m} = 3.9\ \text{Nm}.$$

Belt drives

Driving torque may be transmitted between two parallel shafts by means of a belt passed between pulleys fixed to each shaft. Single or multiple belts may be used and the belts and pulley grooves are usually vee shaped to increase friction and reduce the chance of slipping.

The belt is tightened between the pulleys and driving torque is transmitted by friction between the belt and pulley. There will be a greater tension on one side of the belt than on the other when driving torques are transmitted. These two sides are referred to as the tight side and the slack side, as shown in Figure 6.2. Vee belts do not give a positive drive in the sense that gear drives do, since there will be some load at which the belt will slip and some slip will be present if the belt tension is incorrectly adjusted. Belt drives are made with rubber teeth which combine the advantages of a

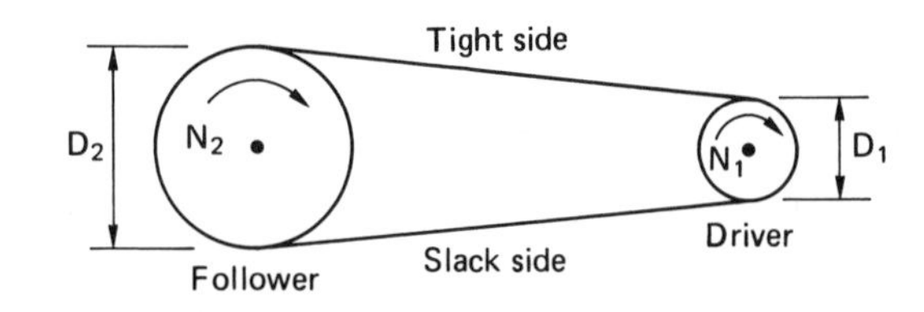

Figure 6.2 Belt drives

positive chain drive coupled with the silence and economy of belt drives.

The pulley connected to the motor or prime mover is called the driver, and the pulley connected to the load the driven or the follower.

Suppose the driver pulley has a diameter D_1 metres and rotates at N_1, revolutions per minute. Its peripheral speed will be πD_1N_1 metres per minute. Similarly, the peripheral speed of the follower pulley will be πD_2N_2 metres per minute. Since both pulleys are joined together by the belt,

$$\pi D_1 N_1 = \pi D_2 N_2$$

$$\therefore \quad \frac{N_1}{N_2} = \frac{D_2}{D_1}.$$

Therefore, the pulley speeds are inversely proportional to their diameters. This means that if the pulley diameter is increased the speed will decrease. Changing the size of a pulley will also change the torque. Let the force exerted at the edge of the driver pulley to be F_1 newtons, and the force at the edge of the follower pulley to be F_2 newtons.

The work done in one revolution on the driver pulley will be $\pi D_1N_1F_1$ joules, and the work done on the follower pulley will be $\pi D_2N_2F_2$ joules. Since both pulleys are joined together by the belt,

$$\pi D_1N_1F_1 = \pi D_2N_2F_2.$$

Now since Torque = Force × radius

$$T_1 = F_1 \times \frac{D_1}{2} \qquad \text{and } T_2 = F_2 \times \frac{D_2}{2}$$

$$\therefore \quad \frac{T_1}{T_2} = \frac{N_2}{N_1} = \frac{D_1}{D_2}.$$

Therefore, pulley torque is proportional to the diameter but inversely proportional to the speed. This means that if a belt and pulley system is used to reduce speed the torque will increase. Conversely, a belt and pulley system used to increase speed will reduce torque.

Example

A 100 mm diameter pulley connected to the motor of a pillar drill rotating at 24 rev/s was connected by a belt drive to a 200 mm diameter pulley connected to the chuck. If the driver pulley provides a torque of 200 Nm find the speed and torque developed at the follower pulley.

Let $D_1 = 100$ mm = diameter of driver
$D_2 = 200$ mm = diameter of follower
$N_1 = 24$ rev/s = speed of driver pulley
N_2 = speed of follower pulley

$$\frac{N_1}{N_2} = \frac{D_2}{D_1}$$

$$\therefore \quad N_2 = \frac{N_1D_1}{D_2} \text{ (rev/s)}$$

$$N_2 = \frac{24 \text{ rev/s} \times 100 \text{ mm}}{200 \text{ mm}} = 12 \text{ rev/s}.$$

Let $T_1 = 200$ Nm = torque developed by driver
T_2 = torque developed at the follower

$$\frac{T_1}{T_2} = \frac{D_1}{D_2}$$

$$\therefore \quad T_2 = \frac{T_1 \times D_2}{D_1} \text{ (Nm)}$$

$$T_2 = \frac{200 \text{ Nm} \times 200 \text{ mm}}{100 \text{ mm}} = 400 \text{ Nm}.$$

Speed cones

A belt driven between pulleys of different diameters provides a convenient method of speed changing, which is widely used in industry. Stepped pulleys are fixed on to parallel shafts as shown in Figure 6.3.

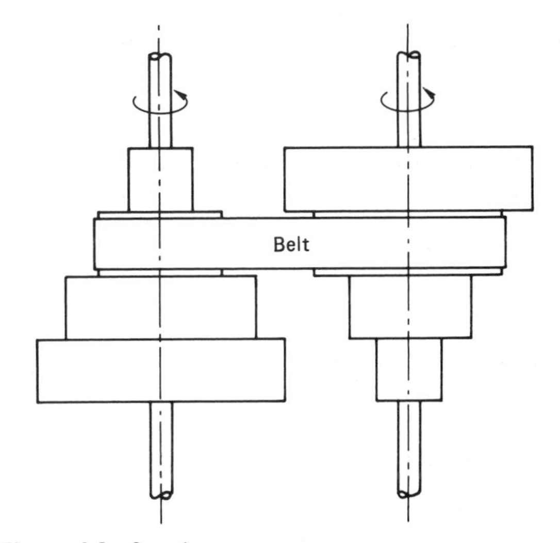

Figure 6.3 Speed cones

The pulleys are arranged so that the sum of the diameters of the pulleys in each section of the two shafts is constant, allowing the same belt length to be used to link each pair of sections. Moving the belt to occupy various positions on the pulleys gives various speed ratios between the shafts. An arrangement such as that shown in Figure 6.3 is known as a pair of speed cones and is used on machine tools such as lathes and pillar drills to give different load speeds for a constant motor speed.

Exercises

1. If we assume the acceleration due to gravity to be 10 m/s^2 a 50 kg bag of cement falling to the ground will exert a force of:
   (a) 5 N, (b) 50 N, (c) 100 N, (d) 500N.
2. The work done by a man carrying a 50 kg bag of cement up a 10 m ladder, assuming the acceleration due to gravity to be 10 m/s^2, will be:
   (a) 50 J, (b) 500 J, (c) 5000 J, (d) 10 000 J.
3. A building hoist is to be used to raise 60 50 kgs bags of cement to the top of a 100 m high building in 1 minute. Assuming the acceleration due to gravity to be 10 m/s^2 the size of the hoist motor would be:
   (a) 10 kW, (b) 50 kW, (c) 60 kW, (d) 100 kW.
4. A passenger lift has the capacity to raise 500 kg at the rate of 2 m/s. Assuming the acceleration due to gravity to be 10 m/s^2, the rating of the lift motor will be:
   (a) 5 kW, (b) 10 kW, (c) 50 kW, (d) 100 kW.
5. Two pulleys are connected by a belt to drive a load. The pulley connected to the motor is 100 mm in diameter and rotates at 20 rps. The load pulley is 75 mm in diameter and, therefore, its speed will be:
   (a) 6.66 rps, (b) 26.66 rps, (c) 200 rps, (d) 375 rps.
6. A driver and driven pulley have diameters of 150 mm and 100 mm respectively. The driver pulley exerts a torque of 100 Nm and therefore the driven pulley will exert a torque of:
   (a) 66.66 Nm, (b) 150 Nm, (c) 1500 Nm, (d) 2660 Nm.
7. Define the terms and give examples of
   (a) potential energy, (b) kinetic energy.
8. Define the conservation of energy theory and give examples.
9. Explain with the aid of a simple sketch the meaning of the term torque as used in electrical and mechanical engineering.
10. Explain how torque varies with speed of rotation.
11. Describe one use of speed cones and explain how they can be used to vary speed and torque.
12. A lift motor is to be used to raise a constant load of 2000 kg at a speed of 0.3 m/s. The motor is supplied at 415 V and works at a p.f. of 0.8 lagging. Find the current taken by the motor, assuming the acceleration due to gravity to be 9.81 m/s^2.

CHAPTER 7

Electrical machines

The foundations of electrical machine principles and construction were laid down in Chapter 10 of *Basic Electrical Installation Work*. In this chapter we will concentrate upon the installation and control of motors, but to begin with let us look at a little more theory which is relevant to electrical machines.

Force on a conductor

If a current carrying conductor is placed into the field of a permanent magnet as shown in Figure 7.1(c) a force F will be exerted on the conductor to push it out of the magnetic field.

To understand the force, let us consider each magnetic field acting alone. Figure 7.1(a) shows the magnetic field due to the current carrying conductor only. Figure 7.1(b) shows the magnetic field due to the permanent magnet in which is placed the conductor carrying no current. Figure 7.1(c) shows the effect of the combined magnetic fields which are distorted and, because lines of magnetic flux never cross, but behave like stretched elastic bands, always trying to find the shorter distance between a north and south pole, the force F is exerted on the conductor, pushing it out of the permanent magnetic field.

This is the basic motor principle, and the force F is dependent upon the strength of the magnetic field B, the magnitude of the current flowing in the conductor I and the length of conductor within the magnetic field l. The following equation expresses this relationship:

force $F = BlI$ (N)

where B is in tesla
l is in metres
I is in amperes
F is in Newtons

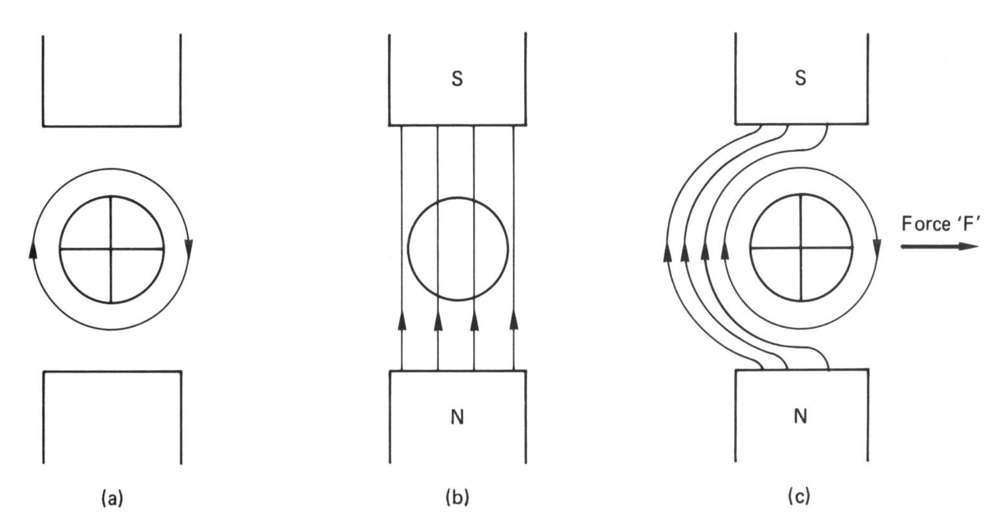

Figure 7.1 Force on a conductor in a magnetic field

Example 1
A coil which is made up of a conductor some 15 m in length, lies at right angles to a magnetic field of strength 5 tesla. Calculate the force on the conductor when 15 A flows in the coil.

$F = BlI$ (N)
$F = 5\ \text{T} \times 15\ \text{m} \times 15\ \text{A} = 1125\ \text{N}$

Production of a rotating magnetic field

If a three phase supply is connected to three separate windings equally distributed around the stationary part or stator of an electrical machine, an alternating current circulates in the coils and establishes a magnetic flux. The magnetic field established by the three phase currents travels in a clockwise direction around the stator, as can be seen by considering the various intervals of time 1 to 6 shown in Figure 7.2. The three phase supply establishes a rotating magnetic flux which rotates at the same speed as the supply frequency. This is called synchronous speed;

$$\text{Synchronous speed} = n_S = \frac{f}{P} \text{ or } N_S = \frac{60f}{P}$$

where n_S is measured in revolutions per second
N_s is measured in revolutions per minute
f is the supply frequency measured in Hz.
P is the number of pole pairs.

Example
Calculate the synchronous speed of a four pole machine connected to a 50 Hz mains supply.

$$n_S = \frac{f}{p} \text{ (rps)}$$

A four pole machine has two pair of poles

$$\therefore \quad n_S = \frac{50\ \text{Hz}}{2} = 25\ \text{rps}$$

$$\text{or } N_S = \frac{60 \times 50\ \text{Hz}}{2} = 1500\ \text{rpm.}$$

This rotating magnetic field is used to practical effect in the induction motor.

Three phase induction motor

When a three phase supply is connected to insulated coils set into slots in the inner surface of the stator or stationary part of an induction motor as shown in Figure 7.3(a), a rotating magnetic flux is produced. The rotating magnetic flux cuts the conductors of the rotor and induces an emf in the rotor conductors by Faraday's Law which states, 'when a conductor cuts or is cut by a magnetic field, an emf is induced in that conductor, the magnitude of which is proportional to the *rate* at which the conductor cuts or is cut by the magnetic flux'. This induced emf causes rotor currents to flow and establish a magnetic flux which reacts with the stator flux and causes a force to be exerted on the rotor conductors, turning the rotor as shown in Figure 7.3(b).

The turning force or torque experienced by the rotor is produced by inducing an emf into the rotor conductors due to the *relative* motion between the conductors and the rotating field. The torque produces rotation in the same direction as the rotating magnetic field. At switch on, the rotor speed increases until it approaches the speed of the rotating magnetic flux, that is, the synchronous speed. The faster the rotor revolves the less will be the difference in speed between the rotor and the rotating magnetic field. By Faraday's Laws, this will result in less induced emf, less rotor current and less torque on the rotor. The rotor can never run at synchronous speed because if it did so, there would be no induced emf, no current and no torque. The induction motor is called an asynchronous motor. In practice the rotor runs at between 2% and 5% below the synchronous speed so that a torque can be maintained on the rotor which overcomes the rotor losses and the applied load.

The difference between the rotor speed and synchronous speed is called slip where

$$\text{Per unit slip } s = \frac{n_S - n}{n_S} = \frac{N_S - N}{N_S}$$

where n_s = synchronous speed in rev/sec
N_s = synchronous speed in rev/min.
n = rotor speed in rev/sec.
N = rotor speed in rev/min.

Percentage slip = per unit slip × 100%

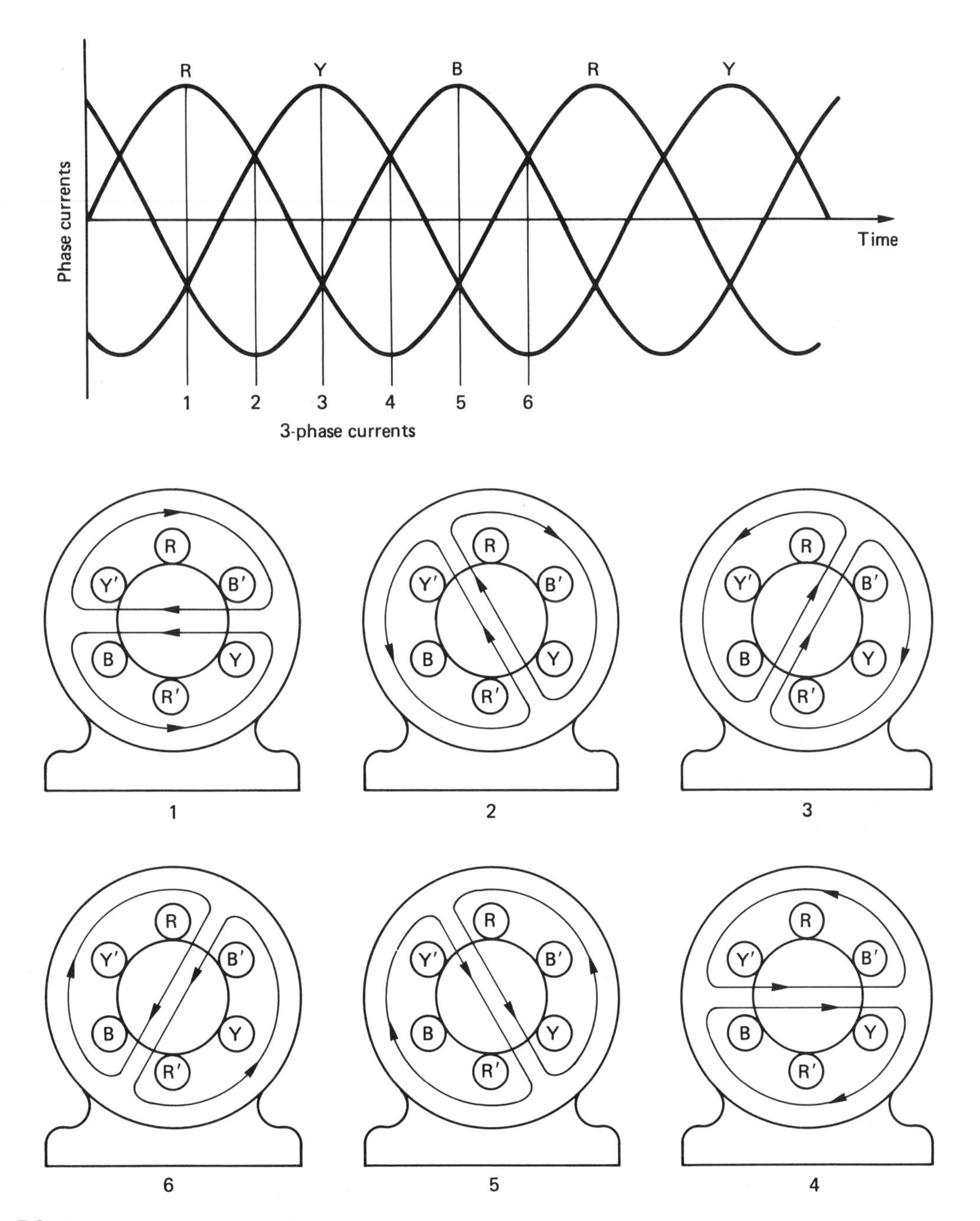

Figure 7.2 Distribution of magnetic flux due to three phase currents

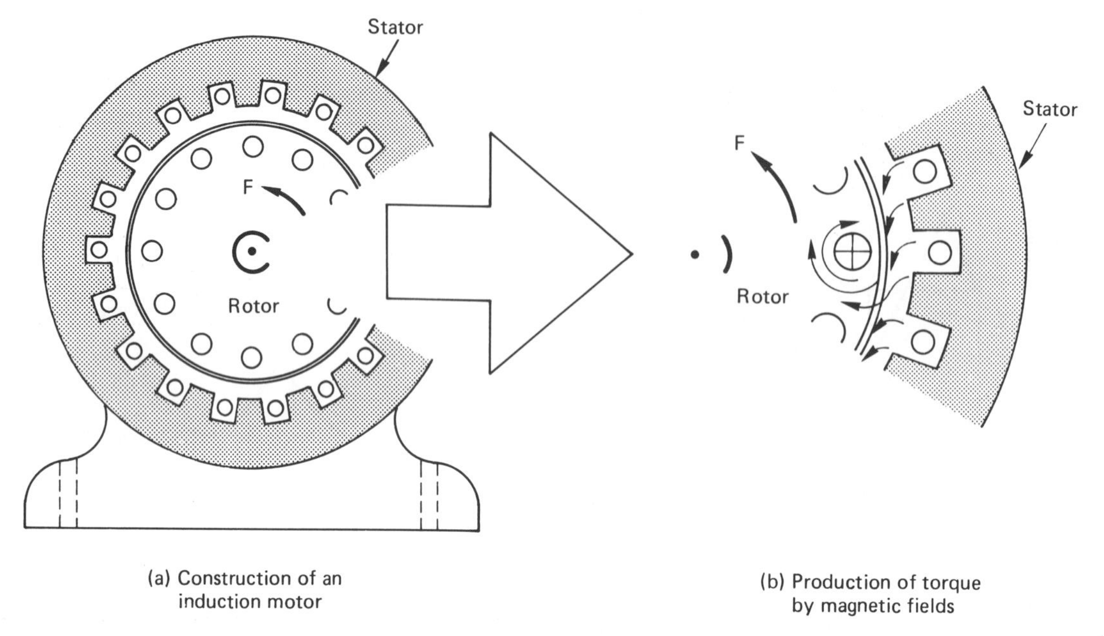

Figure 7.3 Segment taken out of an induction motor to show turning force

Example
A two pole induction motor runs at 2880 rpm when connected to the 50 Hz mains supply. Calculate the per unit slip and the percentage slip.

$$\text{Synchronous speed } N_S = \frac{60 \times f}{p} \text{ (rpm)}$$

$$\therefore \quad N_S = \frac{60 \times 50 \text{ Hz}}{1} = 3000 \text{ rpm.}$$

$$\text{Per unit slip } s = \frac{N_S - N}{N_S}$$

$$\therefore \quad s = \frac{3000 \text{ rpm} - 2880 \text{ rpm}}{3000 \text{ rpm}}$$

$$s = 0.04.$$

$$\text{percentage slip } s = \text{per unit slip} \times 100\%$$

$$s = 0.04 \times 100 = 4\%.$$

Rotor construction

There are two types of induction motor rotor; the wound rotor and the cage rotor. The cage rotor consists of a laminated cylinder of silicon steel with copper or aluminium bars slotted in holes around the circumference and short circuited at each end of the cylinder as shown in Figure 7.4. In small motors the rotor is cast in aluminium. Better starting and quieter running is achieved if the bars are slightly skewed. This type of rotor is extremely robust and since there are no external connections there is no need for slip-rings or brushes. A machine fitted with a cage rotor does suffer from a low starting torque and the machine must be chosen which has a higher starting torque than the load, as shown by curve (b) in Figure 7.5. A machine with the characteristic shown by curve (a) in Figure 7.5 would not start since the load torque is greater than the machine starting torque.

Alternatively the load may be connected after the motor has been run up to full speed, or extra resistance can be added to a wound rotor through slip-rings and brushes since this improves the starting torque, as shown by curve (c) in Figure 7.5. The wound rotor consists of a laminated cylinder of silicon steel with copper coils embedded in slots around the circumference. The windings may be connected in star or delta and the end connections brought out to slip rings mounted on the shaft. Connection by carbon brushes can then be made to an external resistance to improve starting, but once normal running speed is achieved the external resistance is short circuited. Therefore, the principle of operation for both types of rotor is the same.

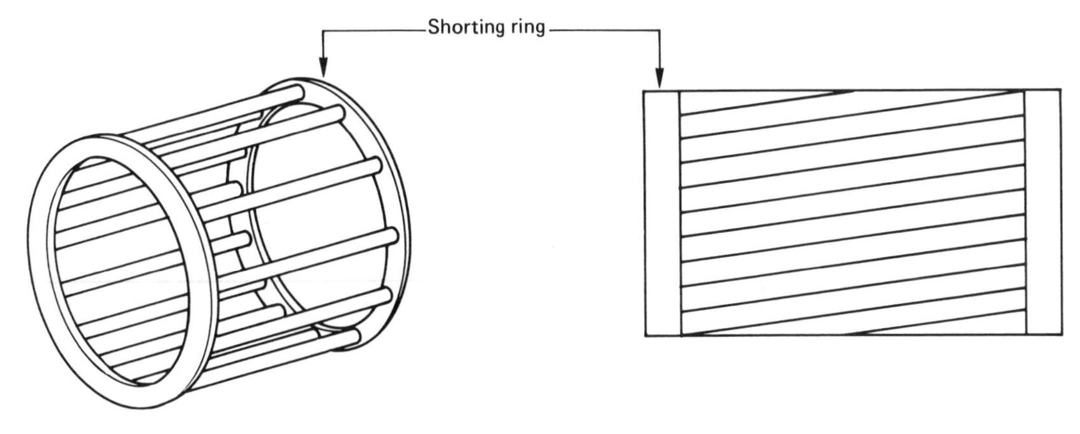

Figure 7.4 Construction of a cage rotor

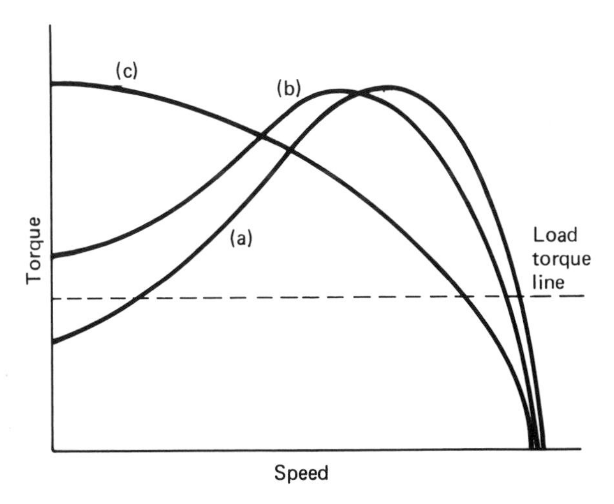

Figure 7.5 Various speed-torque characteristics for an induction motor

Motor starters

The magnetic flux generated in the stator of an induction motor rotates immediately the supply is switched on, and therefore the machine is self-starting. The purpose of the motor starter is not to start the machine, as the name implies, but to reduce heavy starting currents and provide overload and no-volt protection in accordance with the requirements of Regulations 552. All electric motors with a rating above 0.37 kW must be supplied from a suitable motor starter and we will now consider the more common type.

Direct on line starters (d.o.l.)

The d.o.l. starter switches the main supply directly on to the motor. Since motor starting currents can be seven or eight times greater than the running current, the d.o.l. starter is only used for small motors of less than about 5 kW rating.

When the start button is pressed current will flow from the red phase through the control circuit and contactor coil to the blue phase which energises the contactor coil and the contacts close, connecting the three phase supply to the motor, as can be seen in Figure 7.6. If the start button is

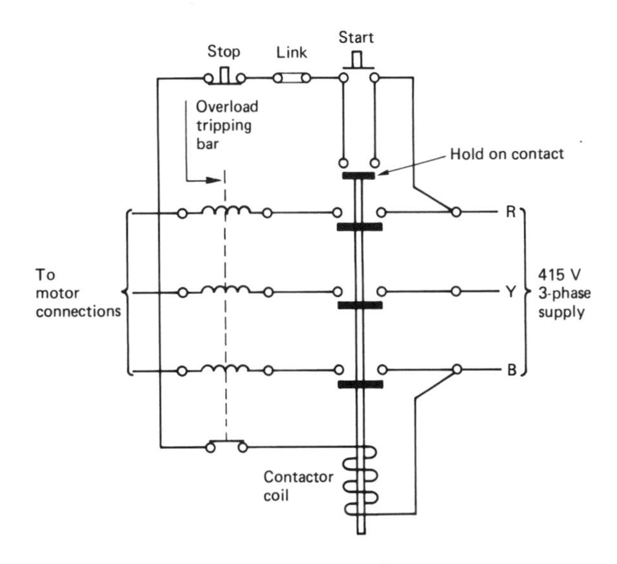

Figure 7.6 Three phase direct on line starter

released the control circuit is maintained by the hold on contact. If the stop button is pressed or the overload coils operate, the control circuit is broken and the contactor drops out, breaking the supply to the load. Once the supply is interrupted the supply to the motor can only be reconnected by pressing the start button. Therefore this type of arrangement also provides no-volt protection.

When large industrial motors have to be started, a way of reducing the excessive starting currents must be found. One method is to connect the motor to a star delta starter.

Star delta starters

When three loads, such as the three windings of a motor are connected in star, the line current has only one third of the value it has when the same load is connected in delta. A starter which can connect the motor windings in star during the initial starting period and then switch to delta connection will reduce the problems of an excessive starting current. This arrangement is shown in Figure 7.7 where the six connections to the three stator phase windings are brought out to the starter. For starting, the motor windings are star connected at the a–b–c end of the winding by the star making contacts. This reduces the phase voltage to about 58% of the running voltage which reduces the current and the motor's torque. Once the motor is running a double throw switch makes the changeover from star starting to delta running, thereby achieving a minimum starting current and maximum running torque. The starter will incorporate overload and no-volt protection, but these are not shown in Figure 7.7 in the interests of showing more clearly the principle of operation.

Auto-transformer starter

An auto-transformer motor starter provides another method of reducing the starting current by reducing the voltage during the initial starting period. Since this also reduces the starting torque, the voltage is only reduced by a sufficient amount to reduce the starting current, being permanently connected to the tapping found to be most appropriate by the installing electrician. Switching the changeover switch to the start position connects the auto-transformer windings in series with the delta-connected motor starter winding. When sufficient speed has been achieved by the motor the changeover switch is moved to the run connections which connect the three phase supply directly on to the motor as shown by Figure 7.8.

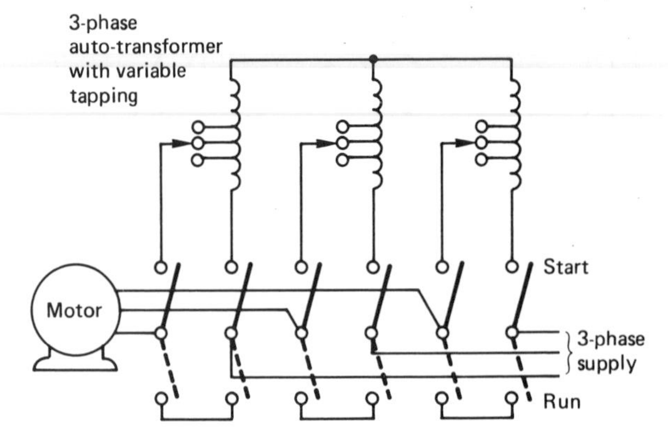

Figure 7.8 Auto-transformer starting

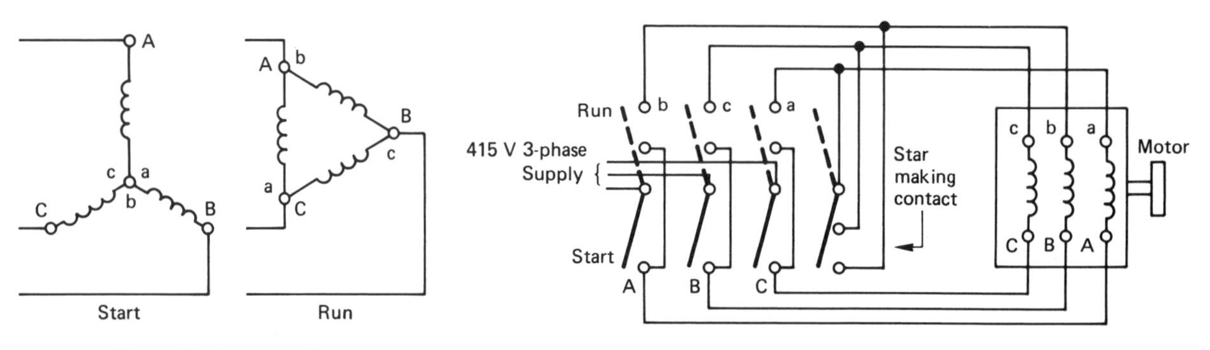

Figure 7.7 Star delta starter

This starting method has the advantage of only requiring three connection conductors between the motor starter and the motor. The starter will incorporate overload and no-volt protection in addition to some method of preventing the motor being switched to the run position while the motor is stopped. These protective devices are not shown in Figure 7.8 in order to show more clearly the principle of operation.

Rotor resistance starter

When starting a machine on load a wound rotor induction motor must generally be used since this allows an external resistance to be connected to the rotor winding through slip rings and brushes, which increases the starting torque as shown in Figure 7.5, curve (c).

When the motor is first switched on the external rotor resistance is at a maximum. As the motor speed increases the resistance is reduced until at full speed the external resistance is completely cut out and the machine runs as a cage induction motor. The starter is provided with overload and no-volt protection and an interlock to prevent the machine being switched on with no rotor resistance connected, but these are not shown in Figure 7.9 since the purpose of the diagram is to show the principle of operation.

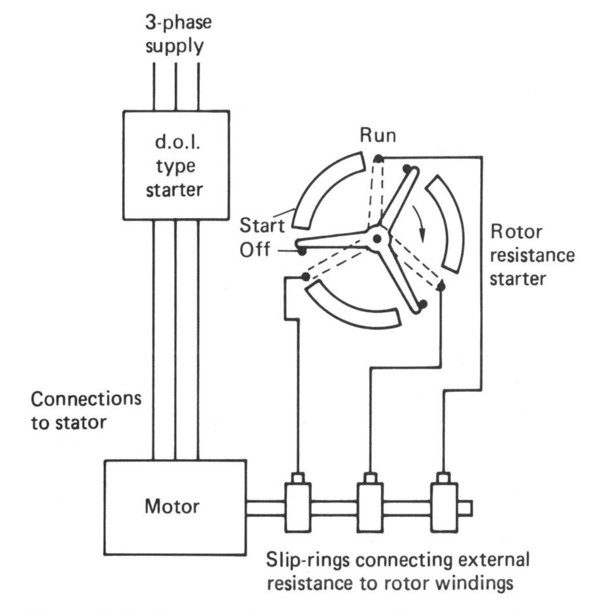

Figure 7.9 Rotor resistance starter for a wound rotor machine

Remote control of motors

When it is required to have stop/start control of a motor at a position other than the starter position, additional start buttons may be connected in parallel and additional stop buttons in series, as shown in Figure 7.10 for the d.o.l. starter. This is the diagram shown in Figure 7.6 with the link removed and a remote stop and start button connected. Additional stop and start facilities are often provided for the safety and convenience of the machine operator.

Installation of motors

Electric motors vibrate when running and should be connected to the electrical installation through a flexible connection. This may also make final adjustments of the motor position easier. For example, to adjust the belt drive tension or align the motor shaft with the final drive shaft. Where the final connection is made with flexible conduit, the tube must not be relied upon to provide a continuous earth path and a separate CPC must be run either inside or outside the flexible conduit (Regulation 543–02–01).

All motors over 0.37 kW rating must be connected to the source of supply through a suitable starter which incorporates overload protection and a device which prevents dangerous restarting of the motor following a mains failure (Regulation 552–01–02 and 03).

The cables supplying the motor must be capable of carrying at least the full load current of the motor (Regulation 552–01–01) and a local means of isolation must be provided to facilitate safe mechanical maintenance (Regulations 476–02–03).

At the supply end, the motor circuit will be protected by a fuse or MCB. The supply protection must be capable of withstanding the motor starting current whilst providing adequate overcurrent protection. There must also be discrimination so that the overcurrent device in the motor starter operates first in the event of an excessive motor current.

Most motors are 'continuously rated'. This is the load at which the motor may be operated continuously without overheating.

Many standard motors have Class A insulation which is suitable for operating in ambient temperatures up to about 55°C. If a Class A motor is

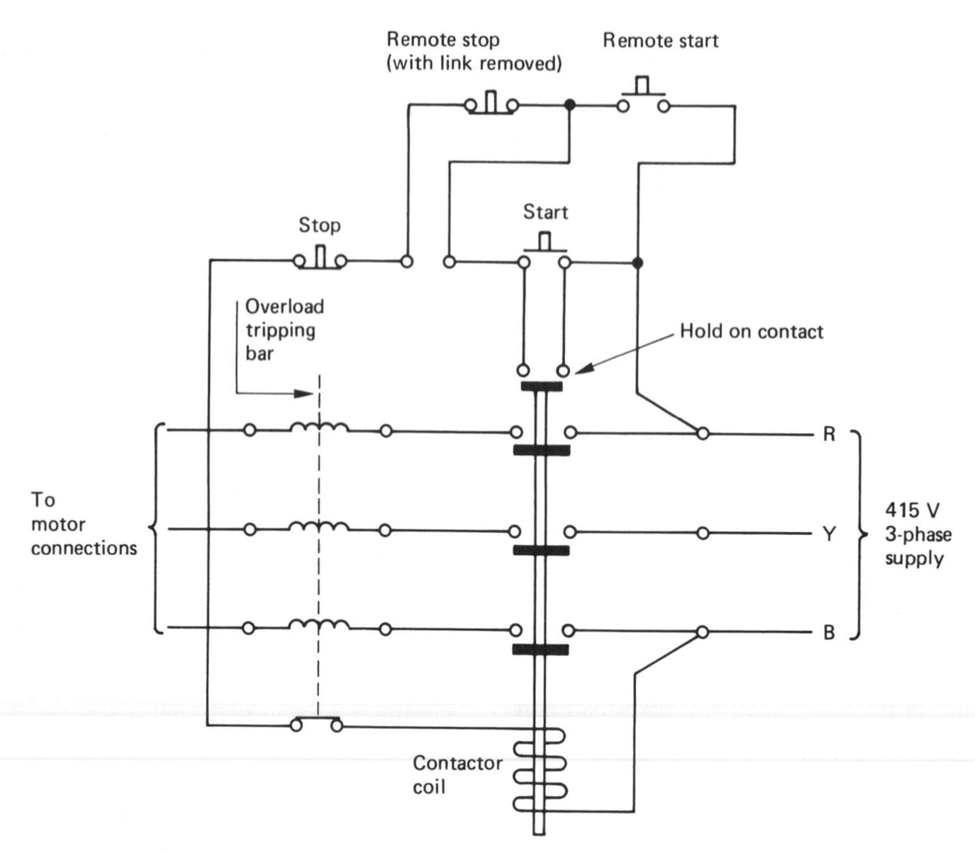

Figure 7.10 Remote stop/start connections to d.o.l. starter

to be operated in a higher ambient temperature, the continuous rating may need to be reduced to prevent damage to the motor. The motor and its enclosure must be suitable for the installed conditions and must additionally prevent anyone coming into contact with the internal live or moving parts. Many different enclosures are used depending upon the atmosphere in which the motor is situated. Clean air, damp conditions, dust particles in the atmosphere, chemical or explosive vapours will determine the type of motor enclosure. In high ambient temperatures it may be necessary to provide additional ventilation to keep the motor cool and prevent the lubricating oil thinning. The following motor enclosures are examples of those to be found in industry:

Screen protected enclosures prevent access to the internal live and moving parts by covering openings in the motor casing with metal screens of perforated metal or wire mesh. Air flow for cooling is not restricted and is usually assisted by a fan mounted internally on the machine shaft. This type of enclosure is shown in Figure 7.11.

A duct ventilated enclosure is used when the air in the room in which the motor is situated is unsuitable for passing through the motor for cooling. For example, when the atmosphere contains dust particles or chemical vapour. In these cases the air is drawn from a clean air zone outside the room in which the machine is installed as shown in Figure 7.11.

A totally enclosed enclosure is one in which the air inside the machine casing has no connection with the air in the room in which it is installed, but it is not necessarily airtight. A fan on the motor shaft inside the casing circulates the air through the windings and cooling is by conduction through the motor casing. To increase the surface area and assist cooling, the casing is surrounded by fins and an externally mounted fan can increase the flow of air over these fins. This type of enclosure is shown in Figure 7.11.

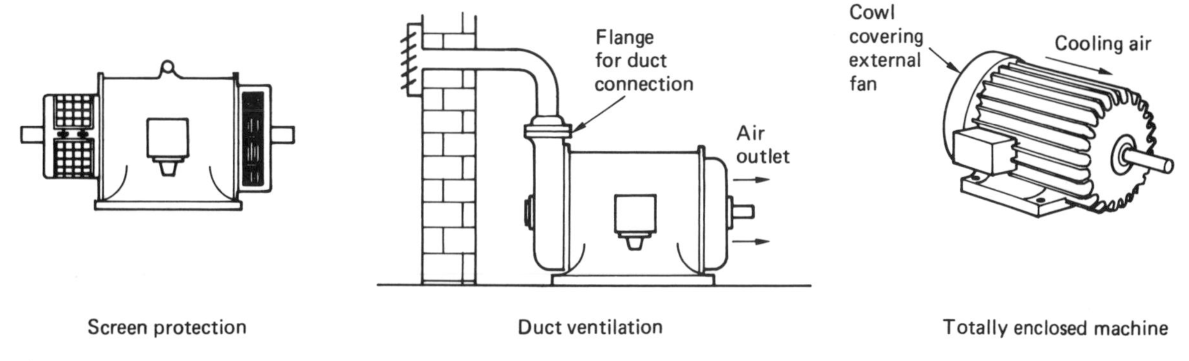

Figure 7.11 Motor enclosures

A flameproof enclosure requires that further modifications be made to the totally enclosed casing to prevent inflammable gases coming into contact with sparks or arcing inside the motor. To ensure that the motor meets the stringent regulations for flameproof enclosures the shaft is usually enclosed in special bearings and the motor connections contained by a wide flange junction box.

When a motor is connected to a load, either by direct coupling or by a vee belt, it is important that the shafts or pulleys are exactly in line. This is usually best achieved by placing a straight edge or steel rule across the flange coupling of a direct drive or across the flat faces of a pair of pulleys as shown in Figure 7.12. Since pulley belts stretch in use it is also important to have some means of adjusting the tension of the vee belt. This is usually achieved by mounting the motor on a pair of slide rails as shown in Figure 7.13. Adjustment is carried out by loosening the motor fixing bolts, screwing in the adjusting bolts which push the motor back, and when the correct belt tension has been achieved the motor fixing bolts are tightened.

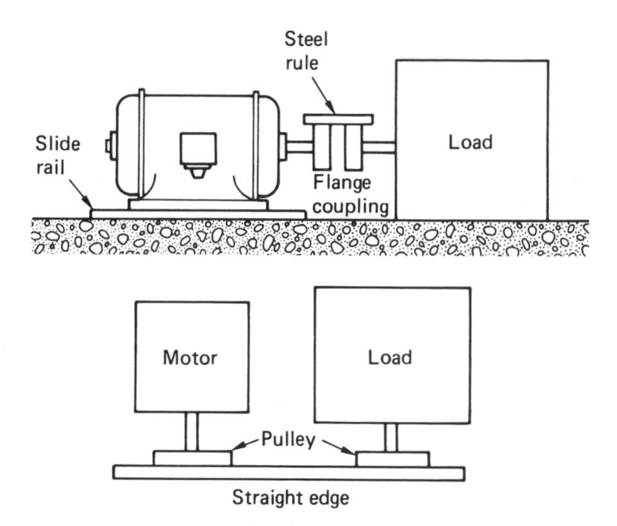

Figure 7.12 Pulley and flange coupling alignment

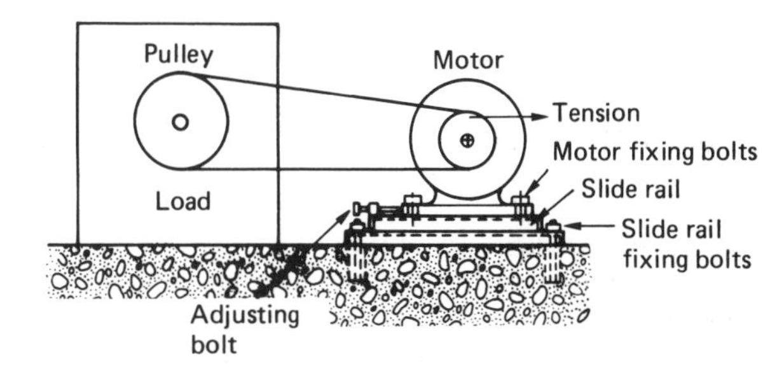

Figure 7.13 Vee belt adjustment of slide rail mounted motor

Power factor correction

Most electrical installations have a low power factor because loads such as motors, transformers and discharge lighting circuits are inductive in nature and cause the current to lag behind the voltage. A capacitor has the opposite effect to an inductor, causing the current to lead the voltage. Therefore, by adding capacitance to an inductive circuit the bad power factor can be corrected. The load current I_L is made up of an in phase component I and a quadrature component I_Q. The

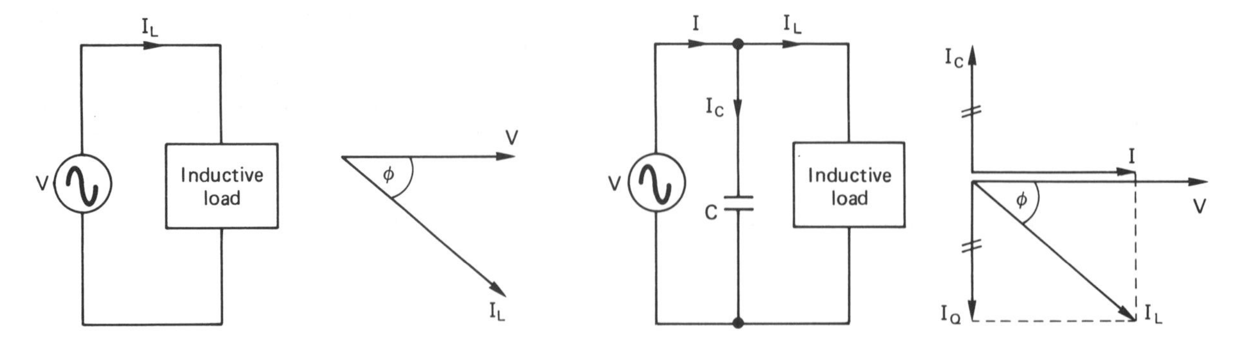

(a) Circuit and phasor diagram for an inductive load with low p.f.

(b) Circuit and phasor diagram for circuit (a) with capacitor correcting p.f. to unity

Figure 7.14 p.f. correction of inductive load

power factor can be corrected to unity when the capacitor current I_C is equal and opposite to the quadrature or reactive current I_Q of the inductive load. The quadrature or reactive current is responsible for setting up the magnetic field in an inductive circuit. Figure 7.14 shows the power factor corrected to unity, that is when $I_Q = I_C$.

A low power factor is considered a disadvantage because a given load takes more current at a low power factor than it does at a high power factor. In Chapter 3 of *Basic Electrical Installation Work* we calculated that a 2 kW load at unity power factor took 8 A but at a bad power factor of 0.4, 20 A were required to supply the same load.

The supply authorities discourage industrial consumers from operating at a bad power factor because

- larger cables and switchgear are necessary to supply a given load,
- larger currents give rise to greater copper losses in transmission cables and transformers,
- larger currents give rise to greater voltage drops in cables,
- larger cables may be required on the consumer's side of the electrical installation to carry the larger currents demanded by a load operating with a bad power factor.

Bad power factors are corrected by connecting a capacitor either across the individual piece of equipment or across the main busbars of the installation. When individual capacitors are used they are usually of the paper dielectric type of construction (see Chapter 2 of *Basic Electrical Installation Work*). This is the type of capacitor used for p.f. correction in a fluorescent luminaire. When large banks of capacitors are required to correct the p.f. of a whole installation paper dielectric capacitors are immersed in an oil tank in a similar type construction to a transformer, and connected on to the main busbars of the electrical installation by suitably insulated and mechanically protected cables.

The current to be carried by the capacitor for p.f. correction and the value of the capacitor may be calculated as shown by the following example.

Example

A 5 kW load with a power factor of 0.7. is connected across a 240 V, 50 HZ supply. Calculate
(a) the current taken by this load,
(b) the capacitor current required to raise the p.f. to unity,
(c) the capacitance of the capacitor required to raise the p.f. to unity.

(a) Since $P = VI \cos \phi$ (W),

$$I = \frac{P}{V \cos \phi} \text{ (A)}$$

$$\therefore \quad I = \frac{5000 \text{ W}}{240 \text{ V} \times 0.7.} = 29.76 \text{ A.}$$

This current lags the voltage by an angle of 45.6 degrees (since $\cos^{-1} 0.7. = 45.6^\circ$) and can therefore be drawn to scale as shown in Figure 7.15. and represented by line AB.

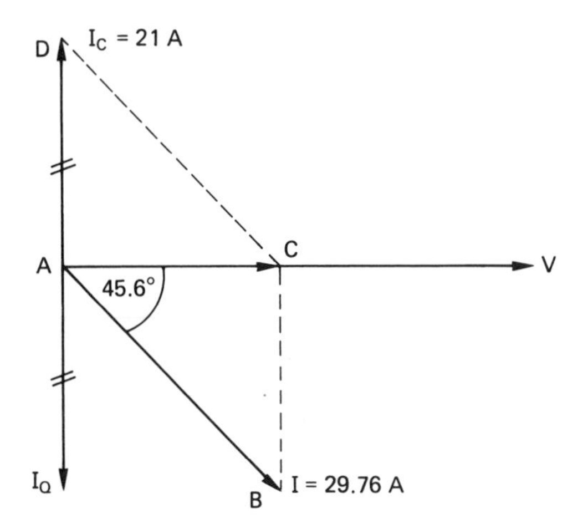

Figure 7.15 Phasor diagram

(b) At unity p.f. the current will be in phase with the voltage, represented by line AC in Figure 7.15. To raise the load current to this value will require a capacitor current I_c which is equal and opposite to the value of the quadrature or reactive component I_Q. The value of I_Q is measured from the phasor diagram and found to be 21 A which is the value of the capacitor current required to raise the p.f. to unit and shown by line AD in Figure 7.15.

(c) Since $I_C = \dfrac{V}{X_C}$ (A), $\quad X_C = \dfrac{V}{I_C}$ (Ω)

$$\therefore \quad X_C = \frac{240\text{ V}}{21\text{ A}} = 11.43\ \Omega.$$

Since $X_C = \dfrac{1}{2\pi fC}$ Ω, $\quad C = \dfrac{1}{2\pi f X_C}$ F.

$$\therefore \quad C = \frac{1}{2 \times \pi \times 50\text{ Hz} \times 11.43\ \Omega} = 278.48\ \mu\text{F}.$$

A 278.48 μF capacitor connected in parallel with the 5 kW load would correct the power factor to unity.

Exercises

1 A coil is made up of 30 m of conductor and is laid within and at right angles to a magnetic field of 4 tesla. The force exerted upon this coil when 5 A flows will be:
(a) 1.875 N, (b) 30 N, (c) 240 N, (d) 600 N.

2 An instrument coil of 20 mm diameter is wound with 100 turns and placed within and at right angles to a magnetic field of flux density 5 tesla. The force exerted on this coil when 15 mA flows in the coil conductors will be:
(a) 0.15 N, (b) 0.47 N, (c) 0.628 N, (d) 1.875 N.

3 The synchronous speed Ns of a 4-pole machine connected to a 50 Hz supply will be:
(a) 200 rpm, (b) 750 rpm, (c) 1500 rpm, (d) 3000 rpm.

4 A four pole induction motor running at 1425 rpm has a percentage slip of:
(a) 2%, (b) 5%, (c) 52.5%, (d) 75%.

5 A laminated cylinder of silicon steel with copper or aluminium bars slotted into holes around the circumference and short circuited at each end of the cylinder is one description of:
(a) a cage rotor, (b) an electro magnet, (c) a linear motor, (d) an induction motor.

6 All electric motors with a rating above 0.37 kW must be supplied with:
(a) protection by MCB, (b) protection by HBC fuses, (c) a motor starter, (d) remote stop/start switches.

7 A star delta starter:
(a) increases the initial starting torque of the motor, (b) reduces the initial starting current of the motor, (c) gives direct connection of the mains voltage to the motor during starting, (d) requires only three connecting conductors between the motor and starter.

8 An auto-transformer starter:
(a) increases the initial starting torque of the motor, (b) increases the initial starting current of the motor, (c) gives direct connection of the mains voltage to the motor during starting, (d) requires only three connecting conductors between the motor and starter.

9 When an electric motor is to be connected to a load via a vee belt, it is recommended that the motor be mounted:

(a) firmly and secured by rawlbolts, (b) firmly and secured on slide rails, (c) loosely and connected by flexible conduit, (d) adjacent to the motor starter.

10 Describe how a rotating magnetic flux produces a turning force on the rotor conductors of an induction motors.

11 Describe with sketches the construction of a wound rotor and cage rotor. State one advantage and one disadvantage for each type of construction.

12 The Regulations require that motor starters incorporate overload protection and no-volt protection. Describe what is meant by overload protection and no-volt protection.

13 Sketch a three phase direct on line motor starter and describe its operation.

14 Sketch the wiring diagram for a three phase direct on line motor starter which incorporates remote stop/start buttons.

15 Sketch the wiring diagram for a star delta motor starter and describe its operation.

16 Sketch the wiring diagram for an auto-transformer motor starter and describe its operation.

17 Sketch the wiring diagram for a rotor resistance motor starter and describe its operation.

18 Use sketches to explain how an electric motor should be installed and connected to a load via a vee belt.

19 Use a block diagram to explain the sequence of control for an electric motor of about 5 kW.

20 Describe what is meant by the 'continuous rating' of a motor.

21 Describe three types of motor enclosure and state one typical application for each type.

22 Use a phasor diagram to explain the meaning of a 'bad power factor'. Describe two methods of correcting the bad power factor due to a number of industrial motors.

23 A 7.5 kW motor with a p.f. of 0.866 is connected to a 415 V 50 Hz supply. Calculate (a) the current taken by the motor, (b) the value of the capacitor required to correct the p.f. to unity.

CHAPTER 8

Instruments and testing

The electrical contractor is charged with a responsibility to carry out a number of tests on an electrical installation and electrical equipment. The individual tests are dealt with in Part 7 of the IEE Regulations and described in Chapter 9 of *Basic Electrical Installation Work*. Electrical measuring instruments are identified by the way in which the instrument movement is deflected in operation. Thus a moving coil instrument movement consists of a coil which is free to move in operation. The basic construction of instruments was described in Chapter 2 of *Basic Electrical Installation Work* but here we will consider the theory of operation, damping and range extension of instruments, making measurements and the testing of an earth electrode.

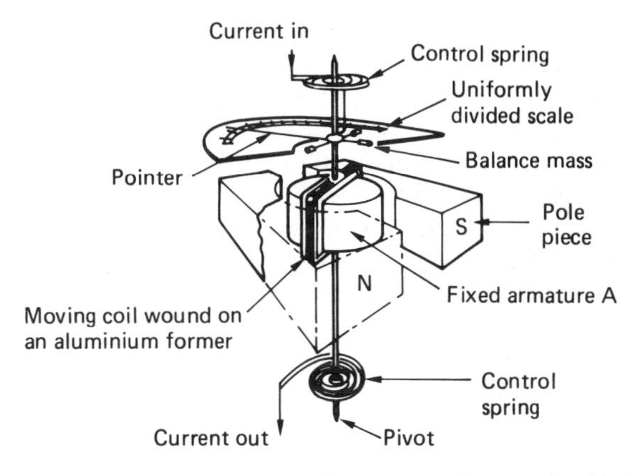

Figure 8.1 Construction of a moving coil meter (by kind permission of G. Waterworth and R. P. Phillips)

Moving coil instruments

A delicate coil is suspended on jewelled bearings between the pole pieces of a strong permanent magnet. The test current is fed to the moving coil through the spiral control springs which also return the pointer to zero after each test reading. The construction is shown in Figure 8.1.

When the test current flows in the moving coil, a magnetic flux is established as shown in Figure 8.2. This flux interacts with the magnetic flux from the permanent magnet and since lines of magnetic flux never cross, the magnetic flux is distorted and a force F is exerted on the coil, rotating it, and moving the pointer across the scale.

The purpose of the soft iron armature A is to establish a uniform magnetic field of equal flux density for the coil's rotation. This gives the MC instrument the qualities of sensitivity and a uniform scale.

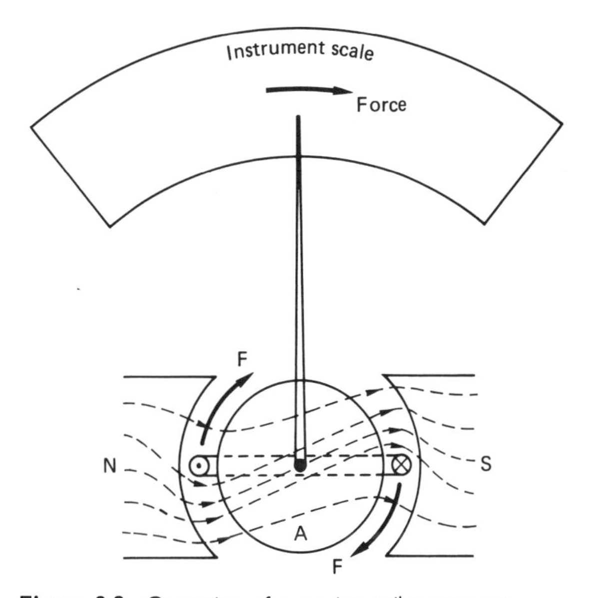

Figure 8.2 Operation of a moving coil movement

The basic moving coil movement will only respond satisfactorily to a d.c. supply since an a.c. supply would reverse the current and magnetic flux in the coil at the supply frequency, resulting in the pointer trembling at some useless mid-point on the scale. To overcome this disadvantage, a.c. test circuits are connected to the MC meter movement through a rectifier circuit as shown in Figure 8.3.

The MC instrument presents a high impedance to the test circuit and draws very little current from it. When used with the rectifier circuit, will give accurate readings over a frequency range from 50 Hz to 50 KHz. This makes a MC instrument suitable for power or electronic circuit measurements. Many commercial multi-range analogue instruments such as the AVO shown in Figure 10.3 use the moving coil movement because of these advantages together with a high sensitivity and linear scale. The wide range of scales achieved by the same meter movement in a commercial instrument is discussed in this chapter under the heading 'Range extension'.

Moving iron instruments

Moving iron instruments operate on basic magnetic principles, either attracting or repelling a piece of soft iron attached to a pointer which moves across a scale. MI instruments can be used to test a.c. or d.c. supplies without the addition of a rectifier circuit.

Attraction type moving iron instrument

The construction of an attraction type moving iron instrument is shown in Figure 8.4. The test current passing through the instrument solenoid establishes a magnetic flux which attracts the soft iron toward the solenoid, moving the pointer across the scale.

The flux density of the magnetic field in which the iron disc moves varies. This creates a non-linear force F on the iron as it moves through the

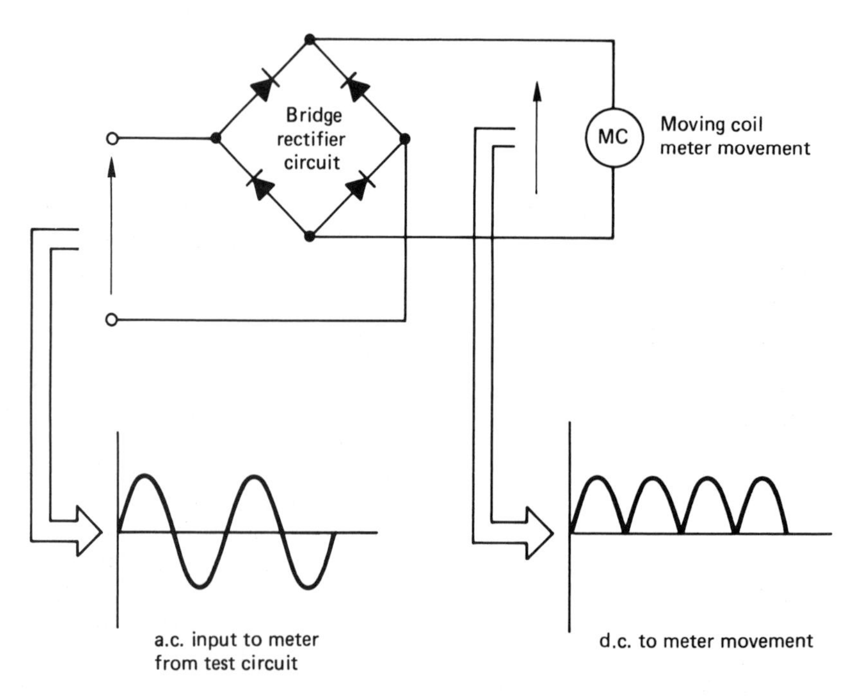

Figure 8.3 Rectified input to moving coil meter movement

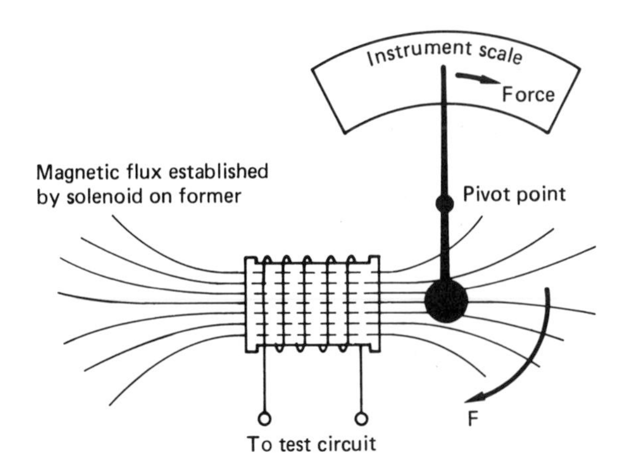

Figure 8.4 Attraction type moving iron instrument

magnetic field and therefore the scale is non-linear. This instrument is usually arranged to have gravity control, by which the force of gravity returns the pointer to zero, and it can only be operated vertically.

Repulsion type moving iron instruments

The repulsion type moving iron instrument is the MI instrument most often found today. The construction is shown in Figure 8.5.

Current passing through the solenoid coil establishes a magnetic field inside the solenoid. This

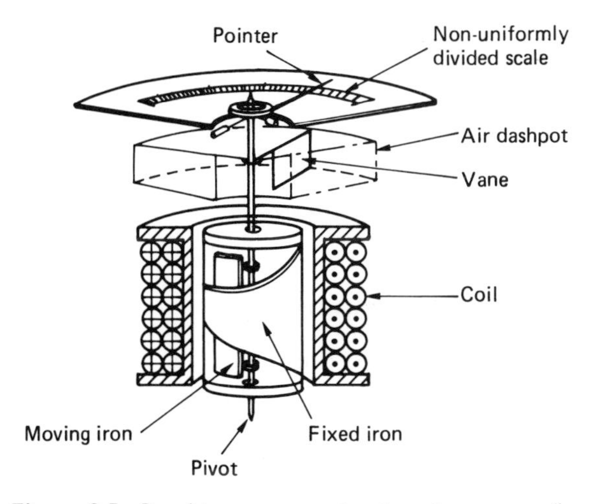

Figure 8.5 Repulsion type moving iron instrument (by kind permission of G. Waterworth and R. P. Phillips)

magnetic field magnetises two pieces of soft iron, one fixed and the other moving, which are close to each other. Since like magnetic poles repel, the moving iron is repelled away from the fixed iron and the pointer is deflected across the scale. The forces exerted between the fixed and moving iron are proportional to the square of the current flowing in the solenoid which results in a non-linear scale.

A linear scale is highly desirable, and to achieve an almost linear scale the manufacturers of commercial MI repulsion type instruments shape the iron pieces. Rectangular moving iron pieces and fixed iron pieces in the shape of a tapered scroll have been found to give good results.

This instrument can usually be operated horizontally since the movement is supported by jewelled bearings and a spiral spring provides the control torque which returns the pointers to zero after each test reading.

Moving iron instruments present a low impedance to the test circuit and will therefore draw current from the test circuit to provide the power necessary for the deflection of the pointer. The frequency range of a MI instrument is only about 50 Hz to 400 Hz and so the applications of moving iron instruments are to be found in power circuits and electrical installations. They are not suitable for electronic circuit measurements.

Damping

If an analogue electrical measuring instrument is correctly connected to a live test circuit, the pointer will rise quite quickly and then settle on the true circuit reading, maintaining a steady value throughout the connection. This effect we take for granted but it can only be achieved if the deflection system is critically damped.

In an undamped system the pointer would rise quickly to the true circuit reading but would then overshoot because of the inertia of the moving system. The forces exerted on the moving system would be insufficient to maintain this higher value and the pointer would fall back toward the true circuit reading, gaining momentum which would take the pointer below the true reading. The forces exerted on the moving system would force the pointer toward the true value, again gaining momentum, resulting in the pointer oscillating about the true value for some time before finally

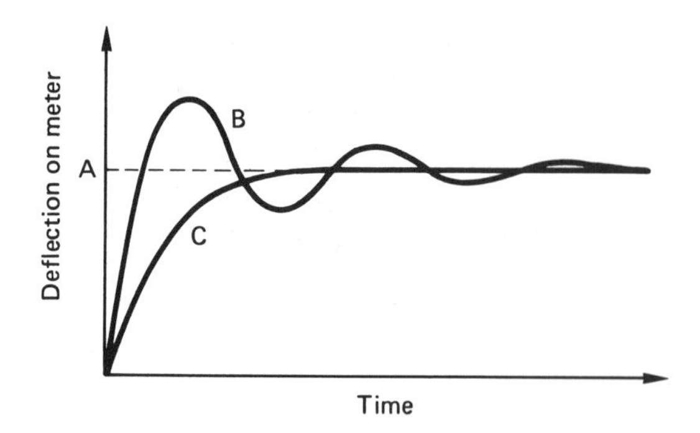

Figure 8.6 Damping curves

coming to rest. This effect is shown in Figure 8.6 where 'A' represents the true reading of the circuit, curve 'B' the undamped system and curve 'C' the correctly or critically damped system. Note that curve 'C', the critically damped system reaches its final steady value much more quickly than the undamped system of curve 'B'.

To achieve damping in an electrical instrument a force is provided which opposes the rise of the moving system toward the final true value. This is achieved in one of three ways; eddy current damping, air valve or air piston damping.

Eddy current damping

When a copper or aluminium disc is rotated between the poles of a permanent magnet, a current is induced into the disc as shown in Figure 8.7. The magnetic field due to this current sets up a force opposing the motion (Lenz's Law) which produces a braking effect. The braking effect only occurs when the disc is in motion and this type of damping is used in moving coil instruments. The delicate moving coil is wound on an aluminium former and as the moving coil moves in the magnetic field so does the aluminium former. Eddy currents are induced in the former which sets up a force to oppose the motion and damping is achieved for the moving coil system.

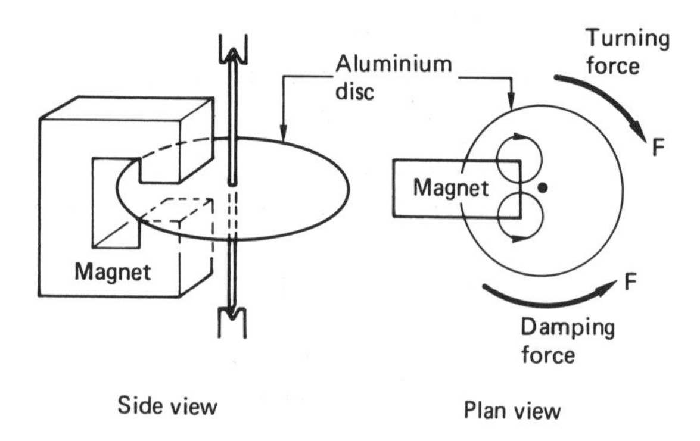

Figure 8.7 Eddy current damping

Air vane damping

This is usually achieved by attaching a thin rectangle of aluminium to the spindle of the instrument movement as shown in Figure 8.8. When the instrument movement deflects, the aluminium rectangle or vane moves in a sector shaped box, compressing the air which exerts a damping force on the vane and on the instrument movement. Air vane damping is used on most moving iron instruments.

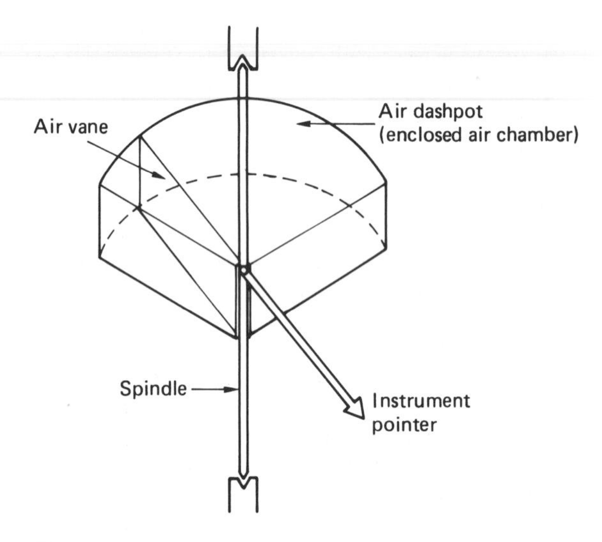

Figure 8.8 Air vane damping

Air piston damping

This is an alternative method of air damping. A piston is attached to the moving system of the instrument as shown in Figure 8.9. When the moving system deflects, the piston is pushed into an air chamber which compresses the air and applies a force which opposes the motion and causes damping of the meter movement.

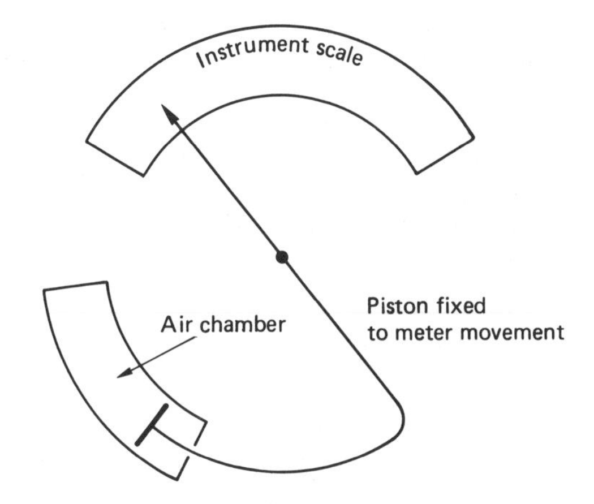

Figure 8.9 Air piston damping

Range extension

Moving iron instruments may be constructed to read 10, 20 or 50 amperes by increasing the thickness and number of solenoid conductors. However, moving coil instruments can only be constructed using a delicate lightweight coil whose maximum current carrying capacity is no more than about 75 mA. To extend the range of moving coil instruments, shunt or series resistors are connected to the test instrument as shown in Figure 8.10.

To extend the range of an ammeter a low resistance shunt resistor is connected across the meter movement. This allows the majority of the circuit current to pass through the shunt and only a very small part of the current to pass through the meter movement. To extend the range of a voltmeter a high resistance series resistor is connected to the meter movement. This limits the current flowing through the meter movement to an acceptable low value.

A multi-range instrument is made up of a number of shunt or series resistors within the instrument which are connected by a range selector switch to form the various scales of the instrument as shown in Figure 8.11.

Example

A moving coil movement has a resistance of 5 Ω and gives full scale deflection when 15 mA flows in the coil. Calculate the value of the resistor which must be connected to the movement in order that the instrument may be used (a) as a 5 A ammeter and (b) as a 100 V voltmeter.

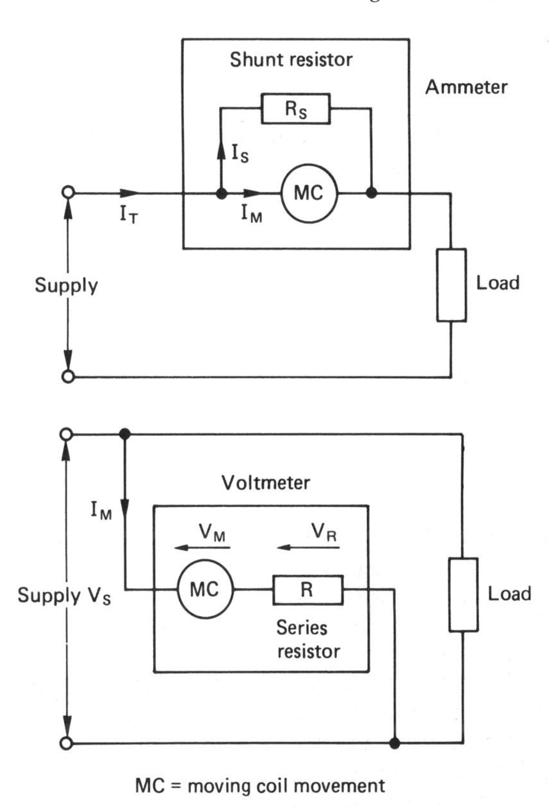

Figure 8.10 Range extension of moving coil movement

For (a), $I_T = I_m + I_S$ (A) (from Figure 8.10)
$I_S = I_T - I_m$ (A)
$I_S = 5\text{ A} - 0.015\text{ A}$
$I_S = 4.985\text{ A}.$

Voltage dropped across the moving coil is given by

$V_m = I_m \times R$ (V)
where R = resistance of MC
$\therefore\quad V_m = 15 \times 10^{-3}\text{ A} \times 5\ \Omega$
$V_m = 75 \times 10^{-3}5\text{ V}.$

Since the shunt resistor is connected in parallel with the moving coil movement,

$V_m = V_S$
where V_S = voltage across the shunt

Shunt resistance $R_S = \dfrac{V_s}{I_s}$ (Ω)

$$\therefore\quad R_S = \frac{75 \times 10^{-3}\text{ V}}{4.985\text{ A}}$$

$R_S = 15.045\text{ m}\Omega.$

For (b), the voltage dropped across the moving coil is given by

$$V_m = I_m \times R \text{ (V)}$$

(where R = resistance of MC)

$$\therefore \quad V_m = 15 \times 10^{-3} \text{ A} \times 5 \ \Omega$$
$$V_m = 75 \times 10^{-3} \text{ V}$$

Supply volts, $V_s = V_m + V_R$ (from Figure 8.10)

$$\therefore \quad V_R = V_S - V_m$$
$$V_R = 100 \text{ V} - 75 \times 10^{-3} \text{ V}$$
$$V_R = 99.925 \text{ V.}$$

Since I_m flows through the MC and the series resistor the resistance of the series resistor

$$R = \frac{V_R}{I_m} \ (\Omega)$$

$$\therefore \quad R = \frac{99.925 \text{ V}}{15 \times 10^{-3} \text{ A}}$$

$$R = 6660 \ \Omega.$$

Making measurements

The electrical contractor is charged by the IEE Regulations for Electrical Installations to test all new installations and major extensions before connection to the mains supply. The contractor may also be called upon to test installations and equipment in order to identify and remove faults. These requirements imply the use of appropriate test instruments, and to take accurate readings consideration should be given to the following points:

- Is the instrument suitable for this test?
- Have the correct scales been selected?
- Is the test instrument correctly connected to the circuit?

Test instruments normally consume a small amount of power in order to provide the torque required to move the pointer across a scale. Most instruments are constructed in a way which makes them highly sensitive, giving full scale deflection with only very small currents, but these currents are drawn from the circuit being tested. When testing power and electrical installation circuits and equipment, the power consumed by the test instrument can often be neglected, but electronic circuits use very little power and therefore when testing electronic circuits a high impedance test instrument must be used.

Example

A 50 Ω load is connected across a 100 V supply. A multi-range meter is correctly connected to read first the current and then the voltage. As an ammeter, the instrument has a resistance of 0.1 Ω and as a voltmeter a resistance of 20 kΩ. The power loss in the instrument for each connection is as follows:

Load current $I = \dfrac{V}{R} = \dfrac{100 \text{ V}}{50 \ \Omega} = 2 \text{ A}$

Power loss in the load

$$= V \times I = 100 \text{ V} \times 2 \text{ A} = 200 \text{ W}$$

Power loss in the ammeter

$$= I^2R = (2 \text{ A})^2 \times 0.1 \ \Omega = 0.4 \text{ W}$$

Power loss in the voltmeter

$$= \frac{V^2}{R} = \frac{(100 \text{ V})^2}{20 \times 10^3 \ \Omega} = 0.5 \text{ W.}$$

This instrument consumes approximately half a watt when taking the readings, which in a power or electrical installation circuit is negligible, but might destroy an electronic circuit.

Many commercial instruments are capable of making more than one test or have a range of scales to choose from. A range selector switch is usually used to choose the appropriate scale as shown in Figure 8.11. A scale range should be chosen which suits the range of the current, voltage or resistance being measured. For example, when taking a reading in the 8 or 9 volt range the obvious scale choice would be one giving 10 V full scale deflection. To make this reading on an instrument with 100 V full scale deflection would lead to errors, because the deflection is too small.

Ammeters must be connected in series with the load and voltmeters in parallel across the load as shown in Figure 8.12. The power in a resistive load may be calculated from the readings of voltage and current since P = VI. This will give accurate calculations on both a.c. and d.c. supplies but when measuring the power of an a.c. circuit which contains inductance or capacitance a wattmeter must be used because the voltage and current will be out of phase.

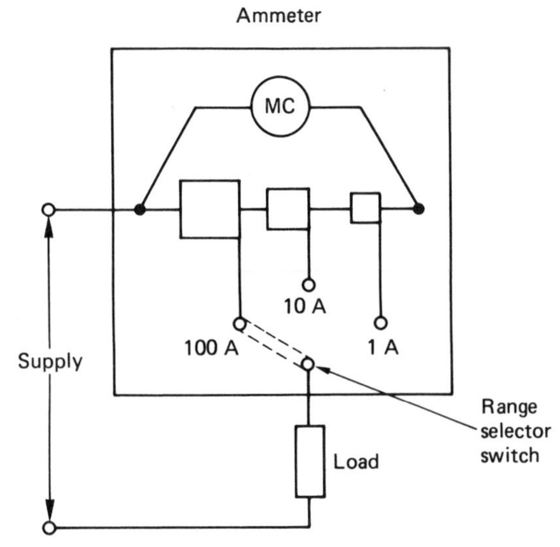

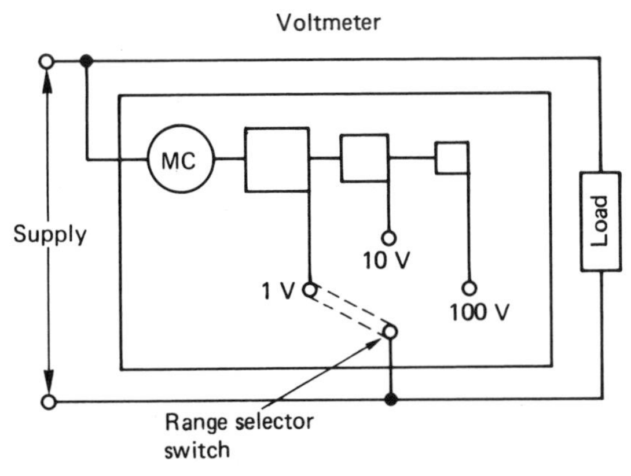

Figure 8.11 The resistor and coil arrangements in a multi-range instrument

Dynamometer wattmeter

A correctly connected wattmeter will give an accurate measure of the power in any a.c. or d.c. circuit. It is essentially a moving coil instrument in which the main magnetic field is produced by two fixed current coils. The moving coil is the voltage

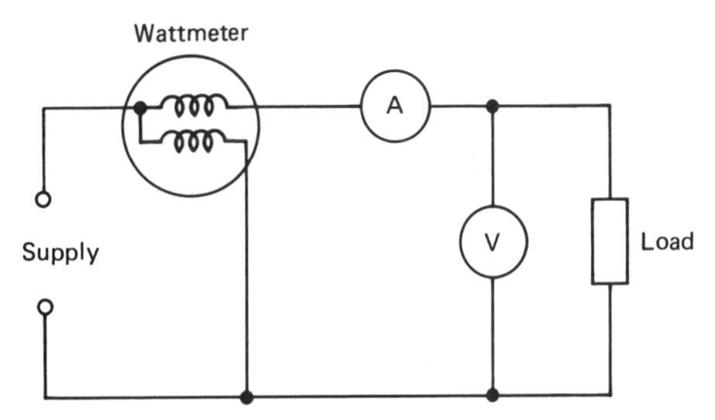

Figure 8.12 Wattmeter, ammeter and voltmeter correctly connected to a load

coil and rotates within the fixed coils being pivoted centrally between them and controlled by spiral hair springs as shown in Figure 8.13.

The main magnetic field is produced by the current in the fixed coil and is proportional to it. The force rotating the moving coil is proportional to its current and the magnetic field strength produced by the fixed coils. The deflection is proportional to the product of the currents in the fixed and moving coils. Since the moving coil current depends upon the voltage and the fixed coils depend upon the current, the meter deflection is proportional to $V \times I$ = power in watts.

Any change in the direction of the current in the circuit affects both coils and the direction of deflection remains unchanged, allowing the instrument to be used on both a.c. and d.c. circuits. On

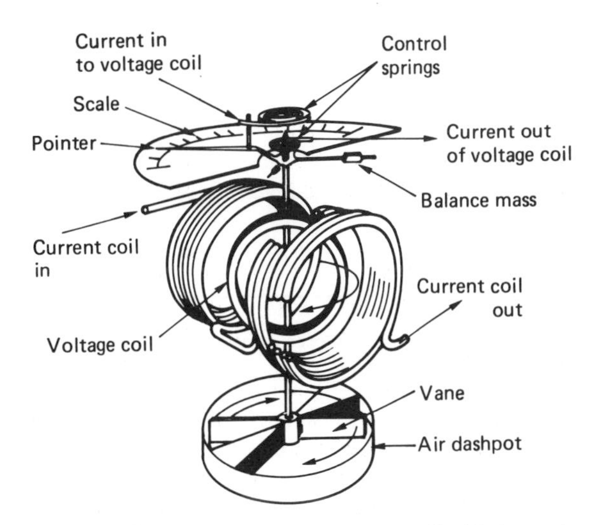

Figure 8.13 A dynamometer wattmeter (by kind permission of G. Waterworth and R. P. Phillips)

a.c. circuits the deflection will be the average value of the product of the instantaneous values of current and voltage, meaning that the wattmeter will measure the true power or active power in the circuit, in which the deflection is proportional to VI cos θ (watts). Damping is achieved by an air vane moving in a dashpot.

Measurement of power in a three phase circuit

One wattmeter method

When three phase loads are balanced, for example in motor circuits, one wattmeter may be connected into any phase as shown in Figure 8.14. This wattmeter will indicate the power in that phase and since the load is balanced the total power in the three phase circit will be given by:

Total power = 3 × Wattmeter reading

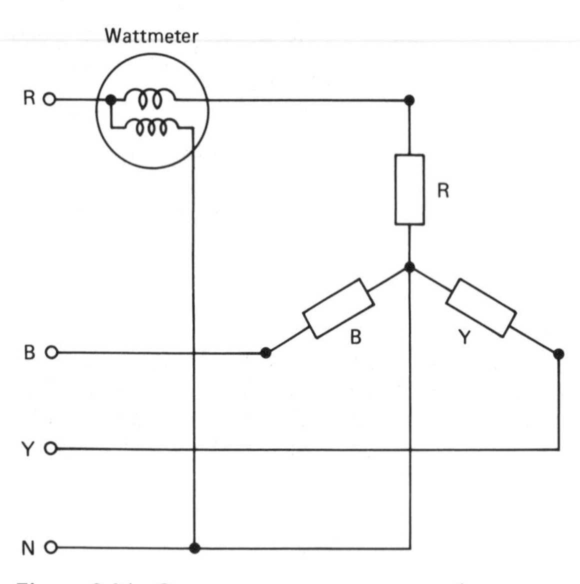

Figure 8.14 One wattmeter measurement of power

Two wattmeter method

This is the most commonly used method for measuring power in a three phase three wire system since it can be used for both balanced and unbalanced loads connected in either star or delta. The current coils are connected to any two of the lines, and the voltage coils are connected to the other line, the one without a current coil connection as shown in Figure 8.15.

Total power = $W_1 + W_2$

This equation is true for any three phase load, balanced or unbalanced, star or delta connection, provided there is no fourth wire in the system.

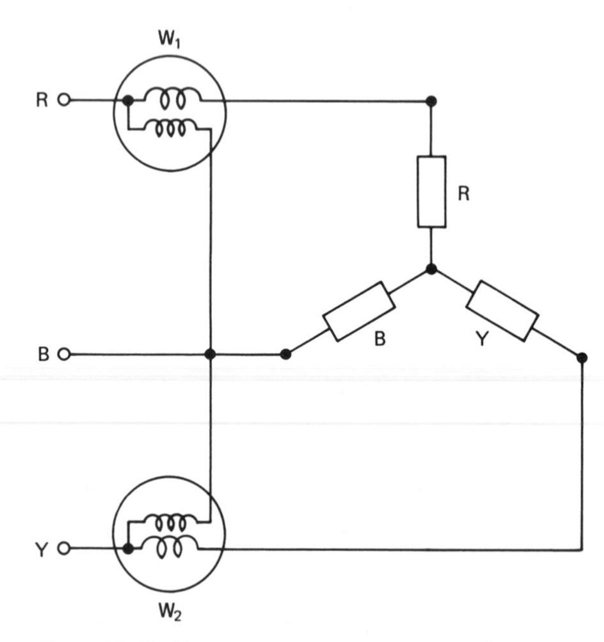

Figure 8.15 Two wattmeter measurement of power

Three wattmeter method

If the installation is four wire, and the load on each phase is unbalanced, then three wattmeter readings are necessary, connected as shown in Figure 8.16. Each wattmeter measures the power in one phase and the total power will be given by

Total power = $W_1 + W_2 + W_3$.

Extension of wattmeter range

The range of a wattmeter can be extended on a.c. circuits by using instrument transformers. A voltage transformer VT is connected to the moving coil circuit of the wattmeter and a current transformer CT is connected to the fixed coils. The VT

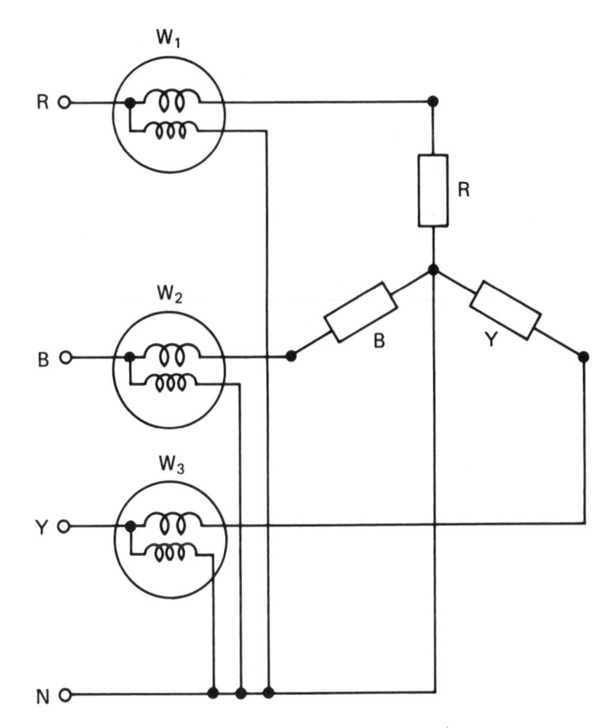

Figure 8.16 Three wattmeter measurement of power

and CT multiplying factors are then applied to the wattmeter readings so that:

$$\text{True power} = \text{Wattmeter reading} \times \text{VT multiplier} \times \text{CT multiplier (W)}$$

An instrument transformer primary carries the current or voltage to be measured and the measuring instrument is connected to the secondary. In this way the instrument measures a small current or voltage which is proportional to the main current or voltage.

The advantages of using instrument transformers are:

- the secondary side of the instrument transformer is wound for low voltage which simplifies the insulation of the measuring instrument and makes it safe to handle,
- the transformer isolates the instrument from the main circuit,
- the measuring instrument can be read in a remote, convenient position connected by long leads to the instrument transformer,
- the secondary voltage or current can be standardised (usually 110 V and 5 A), which simplifies instrument changes.

Voltage transformers (VT)

The construction of a VT is similar to the power transformer dealt with in Chapter 3 of *Basic Electrical Installation Work*. The secondary winding of the VT is connected to a voltmeter or the voltage coil of a wattmeter as shown in Figure 8.17. Voltage transformers are operated as a step down transformer, the secondary voltage usually being standardised at 110 V. A large number of turns are wound on the primary and a few on the secondary since

$$\frac{V_P}{V_S} = \frac{N_P}{N_S}.$$

The voltmeter reading must be multiplied by the turns ratio to determine the load voltage.

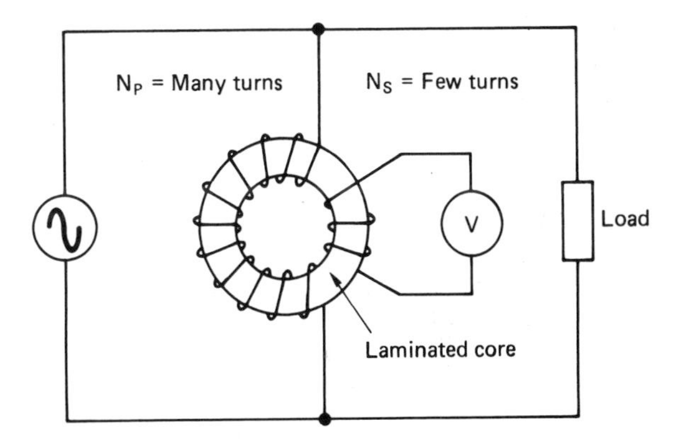

Figure 8.17 A voltage transformer (VT)

Example 1

A voltmeter is connected to 50 turns on the secondary winding of a VT. The primary winding of 250 turns is connected to the main supply. Calculate the supply voltage if the voltmeter reading was 83 V.

$$\text{Primary voltage } V_P = \frac{N_P}{N_s} \times V_S$$

$$\therefore \quad V_P = \frac{250\text{ T}}{50\text{ T}} \times 83\text{ V}$$

$$V_P = 415\text{ V}.$$

As an alternative solution we could say the turns ratio is 250 : 50, that is 5 : 1, and therefore the supply voltage is $5 \times 83 = 415$ V.

Example 2
An electrical contractor wishes to monitor a 660 V supply with a standard 110 V voltmeter. Determine the turns ratio of the VT to perform this task.

$$\frac{V_P}{V_s} = \frac{N_P}{N_s}$$

$$\frac{660\ \text{V}}{110\ \text{V}} = \frac{N_P}{N_s} = \frac{6}{1}.$$

The turns ratio is 6:1. This means that the number of turns on the primary side of the VT must be six times greater than the number of turns on the secondary, which is connected to the 110 V voltmeter.

Current transformers (CT)

The operation of a CT is different to a power transformer although the transformer principle dealt with in Chapter 3 of *Basic Electrical Installation Work* remains the same.

The secondary winding of the CT consists of a large number of turns connected to an ammeter or the current coil of a wattmeter as shown in Figure 8.18(a). The ammeter is usually standardised at 1 A or 5 A and the transformer ratio chosen so that 1 A or 5 A flows when the main circuit carries full load current calculated from the transformer turns ratio

$$\frac{V_P}{V_S} = \frac{I_S}{I_P}.$$

The primary winding is wound with only a few turns and when heavy currents are being measured one turn on the primary may be sufficient. In this case the conductor carrying the main current or the main busbar is passed through the centre of the CT as shown in Figure 8.18(b). This is called a bar primary CT.

Example
An ammeter having a full scale deflection of 5 A is used to measure a line current of 200 A. If the primary is wound with two turns calculate the number of secondary turns required to give full scale deflection.

$$\frac{N_P}{N_S} = \frac{I_S}{I_P}$$

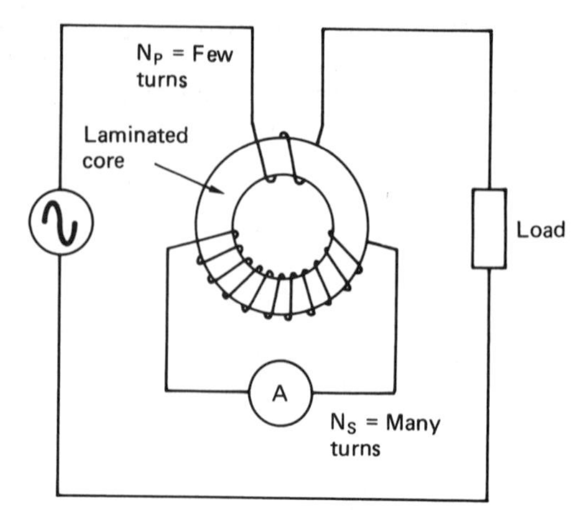

(a) Wound primary current transformer

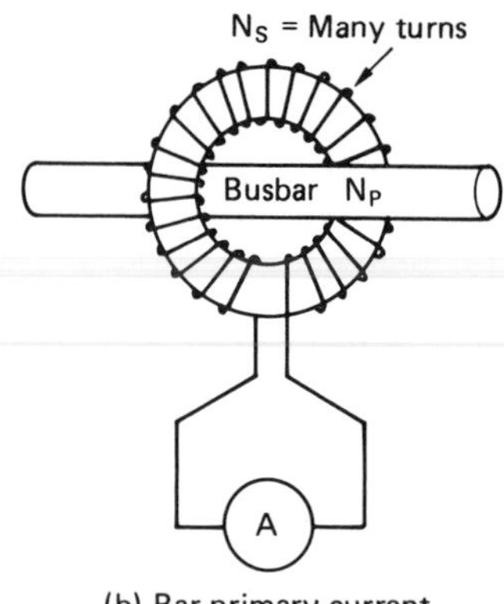

(b) Bar primary current transformer

Figure 8.18 A current transformer (CT)

$$N_S = \frac{N_P \times I_P}{I_S}$$

$$N_S = \frac{2\ \text{T} \times 200\ \text{A}}{5\ \text{A}} = 80\ \text{turns}.$$

With a power transformer a secondary load is necessary to cause a primary current to flow which maintains the magnetic flux in the core at a constant value. With a CT the primary current is the main circuit current and will flow whether the secondary is connected or not.

However, the secondary current through the ammeter is necessary to stabilise the magnetic flux in the core, and if the ammeter is removed the voltage across the secondary terminals could reach a dangerously high value and cause the insulation to break down or cause excessive heating of the core. The CT must never be operated with the secondary terminals open circuited and overload protection should not be provided in the secondary circuit (Regulation 473–01–03). If the ammeter must be removed from the CT then the terminals must first be short circuited. This will not damage the CT and will prevent a dangerous situation arising. The rating of an instrument transformer is measured in volt amperes and called the burden. To reduce errors, the ammeter or voltmeter connected to the CT or VT should be operated at the rated burden.

Example
To determine the power taken by a single phase motor a wattmeter is connected to the circuit through a CT and VT. The test readings obtained were:

Wattmeter reading = 300 W
Voltage transformer turns ratio = 440/110 V
Current transformer turns ratio = 150/5 A

Sketch the circuit arrangements and calculate the power taken by the motor.

The circuit arrangements are shown in Figure 8.19.

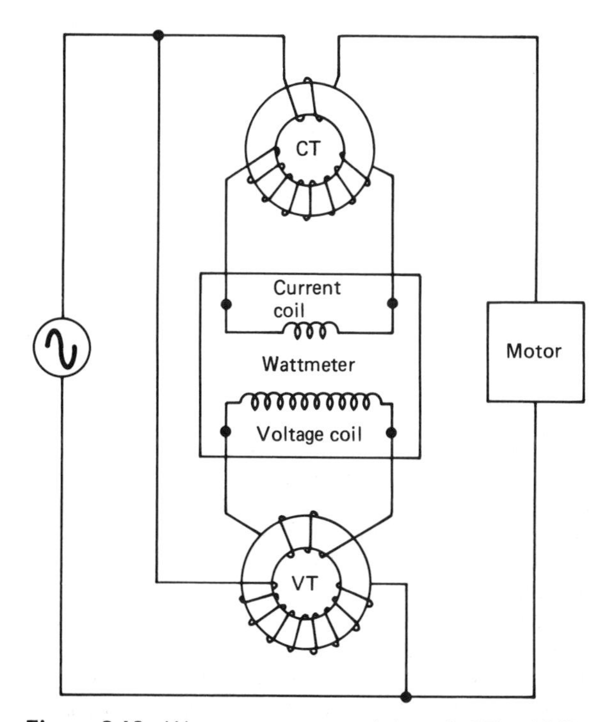

Figure 8.19 Wattmeter connected through CT and VT

Voltage transformer ratio = 440/110 = 4 : 1
Current transformer ratio = 150/5 = 30 : 1

$$\text{True Power} = \text{Wattmeter reading} \times \text{VT multiplier} \times \text{CT multiplier} \quad \text{(W)}$$

True power = 300 W × 4 × 30
True power = 36 kW.

Continuity tester

The construction of the deflection system of a continuity tester is very similar to the moving coil system except that two coils are fixed together at right angles within a permanent magnetic field, as shown in Figure 8.20(a). When measuring the low resistance of a conductor I_1 will be much greater than I_2 or I_3, but if the conductor resistance increases, more current will flow through I_2 and I_3

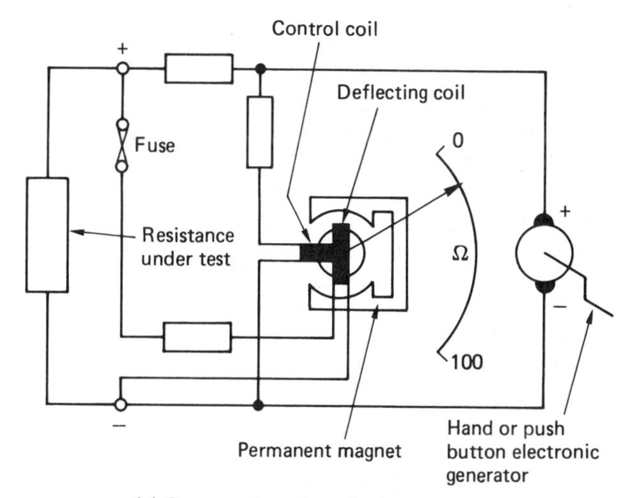

(a) Construction of continuity tester

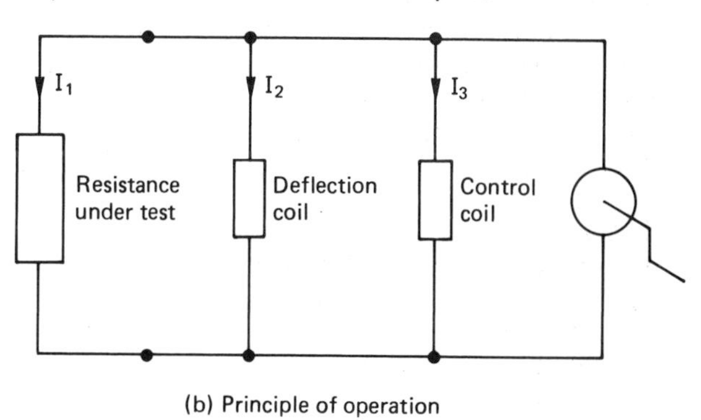

(b) Principle of operation

Figure 8.20 Construction and principle of operation of a continuity tester

and the magnetic fields due to I_2 and I_3 will deflect the pointer across the scale in proportion to the magnitude of these currents shown in Figure 8.20(b).

Insulation resistance tester

The construction of an insulation resistance tester is shown in Figure 8.21. The similarity with Figure 8.20 is because the test instrument for resistance and continuity is combined in the 'megger' instrument.

The insulation resistance should ideally be extremely high in the MΩ range and approaching infinity. The only current in the circuit will therefore flow through the control coil which is designed to set the pointer to infinity. If the resistance decreases because of faulty insulation, the deflecting coil will also carry a current. The magnetic field produced by this current will interact with the main field and deflect the pointer across the scale in proportion to the magnitude of the current.

IEE Regulation 713–04–04 demands that the test voltage indicated in Table 71A be used, that is 500 V for a 240 V and 415 V installation. To achieve this high voltage the instrument incorporates a hand generator or an electronic amplifier operated by a push button.

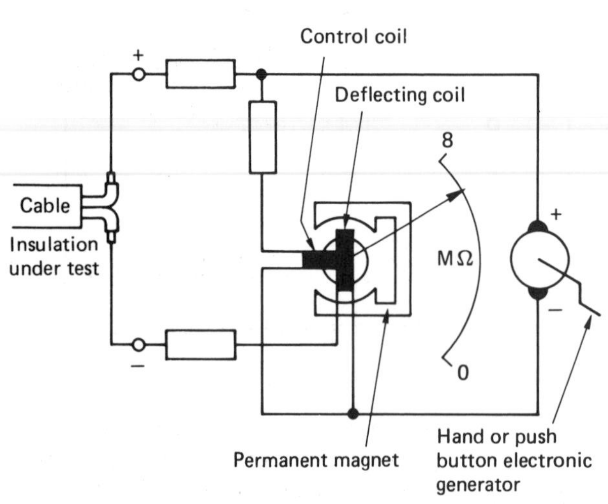

(a) Construction of insulation resistance tester

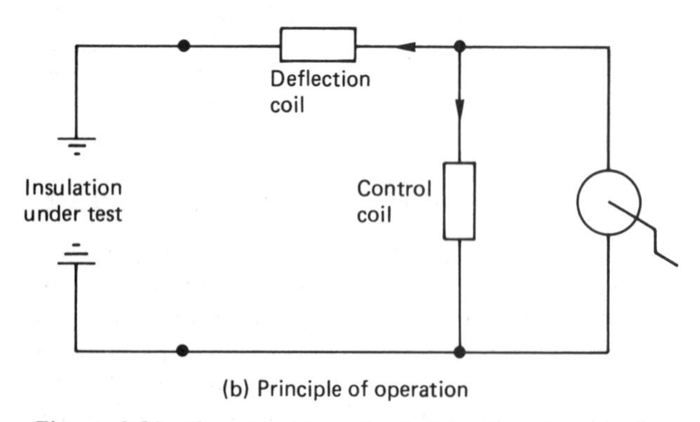

(b) Principle of operation

Figure 8.21 Construction and principle of operation of an insulated resistance tester

Energy meter

The current and voltage coils are wound on the two magnets as shown in Figure 8.22. The current coil establishes a flux Φ_I which is proportional to the current, and the voltage coil establishes a magnetic flux Φ_V.

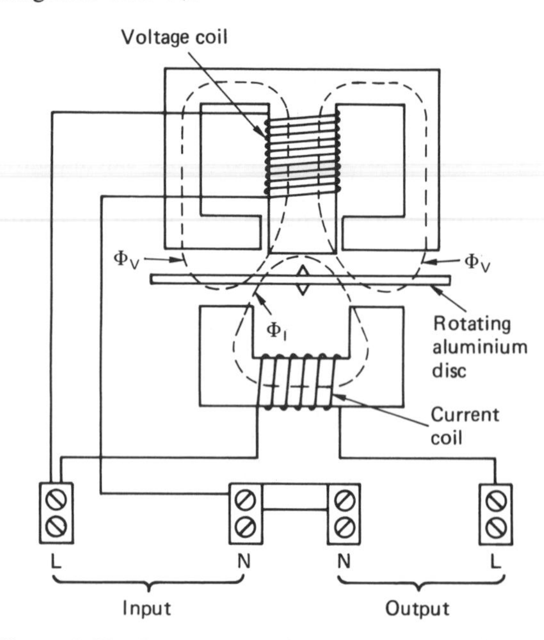

Figure 8.22 Construction of an energy meter

The rotation of the aluminium disc is due to the interaction of these magnetic fields. The magnetic flux establishes eddy currents in the disc which produce a turning force. The force exerted is proportional to the phase angle between the voltage and current coil fluxes; maximum force occurs when they are 90° out of phase. This force is proportional to the true power VI cos θ, which is equal to the speed of rotation of the disc. The number of revolutions in a given time will give a measure of energy since energy = power × time.

The rotating disc spindle is attached through suitable gearing to a revolution counter which is calibrated to read kilowatt hours (kWh), which is the Board of Trade unit of electric energy.

Tong tester

The tong tester or clip-on ammeter works on the same principle as the bar primary current transformer shown in Figure 8.18. The laminated core of the transformer can be opened and passed over the busbar or single core cable. In this way a measurement of the current being carried can be made without disconnection of the supply. The construction is shown in Figure 8.23.

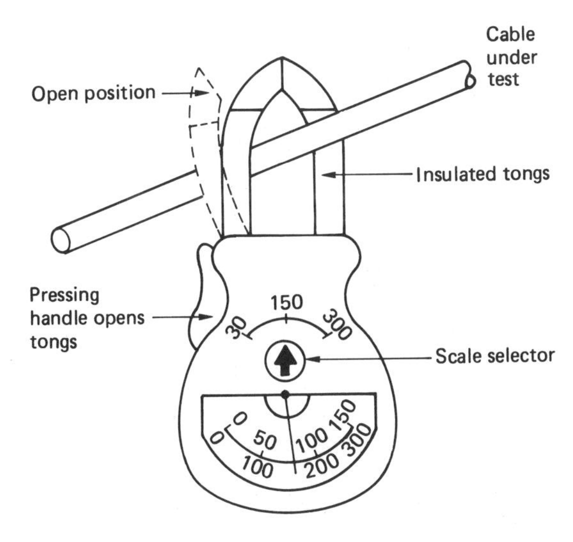

Figure 8.23 Tong tester or clip-on ammeter

Phase sequence testers

Phase sequence is the order in which each phase of a three phase supply reaches its maximum value. The normal phase sequence for a three phase supply is R–Y–B which means that first red, then yellow and finally the blue phase reaches its maximum values.

Phase sequence has an important application in the connection of three phase transformers. The secondary terminals of a three phase transformer must not be connected in parallel until the phase sequence is the same.

A phase sequence tester can be an indicator which is, in effect, a miniature induction motor, with three clearly colour coded connection leads. A rotating disc with a pointed arrow shows the normal rotation for phase sequence R–Y–B. If the sequence is reversed the disc rotates in the opposite direction to the arrow. However, an on-site phase sequence tester can be made by connecting four 240 V by 100 W lamps and a p.f. correction capacitor from a fluorescent luminaire as shown in Figure 8.24.

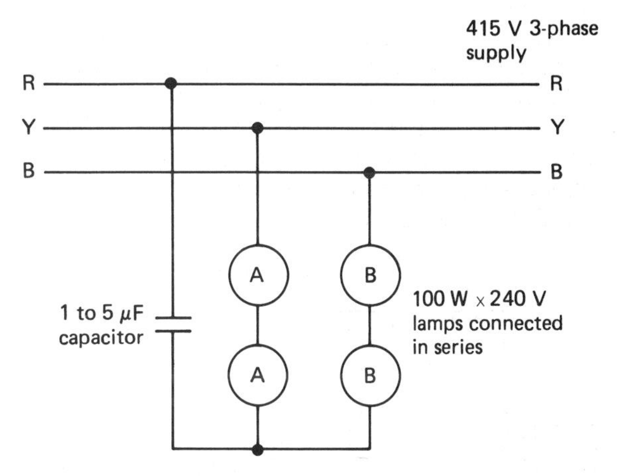

Figure 8.24 Phase sequence test by the lamps bright, lamps dim method

The capacitor takes a leading current which results in a phase displacement in the other two phases. The phasor addition of the voltage in the circuit results in one pair of lamps illuminating brightly whilst the other pair are illuminated dimly. Two lamps must be connected in series as shown in Figure 8.24 because most of the line voltage will be across them during the test.

Earth electrode resistance testing

The general mass of earth can be considered as a large conductor which is at zero potential. Connection to this mass through earth electrodes provides a reference point from which all other voltage levels can be measured. This is a technique which has been used for a long time in power distribution systems.

The resistance to earth of an electrode will depend upon its shape, size and the resistance of the soil. Earth rods form the most efficient electrodes. A rod of about 1 m long will have an earth electrode resistance of between 10 Ω and 200 Ω. Even in bad earthing conditions a rod of about 2 m will normally have an earth electrode resistance which is less than 500 Ω in the United Kingdom. In countries which experience long dry periods of weather the earth electrode resistance may be as high as thousands of ohms.

In the past, electrical engineers used the metal pipes of water mains as an earth electrode, but the recent increase in the use of PVC pipe for water mains now prevents the use of water pipes as the only means of earthing in the United Kingdom, although this practice is still permitted in some countries. The IEE Regulation 542–02–01 recognises the use of the following types of earth electrodes:

Earth rods or pipes
Earth tapes or wires
Earth plates
Earth electrodes embedded in foundations
Metallic reinforcement of concrete structures
Metal pipes
Lead sheaths or other metallic coverings of cables

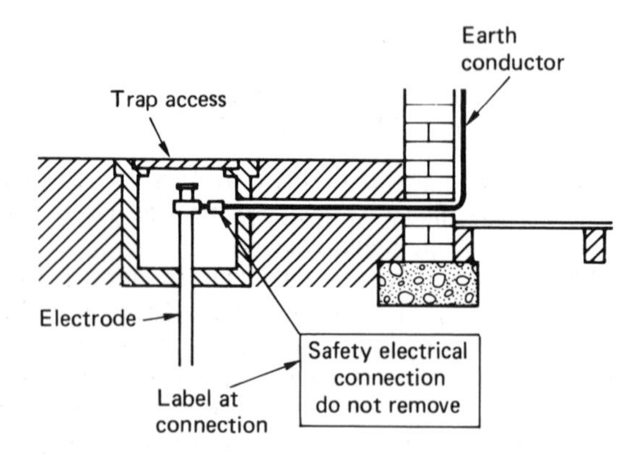

Figure 8.25 Termination of an earth electrode

The earth electrode is sunk in the ground but the point of connection should remain accessible (Regulation 542–04–02). The connection of the earthing conductor to the earth electrode must be securely made with a copper conductor complying with Table 54A and Regulation 542–03–03 as shown in Figure 8.25.

The installation site must be chosen so that the resistance of the earth electrode does not increase above the required value due to climatic conditions such as the soil drying out or freezing, or from the effects of corrosion (542–02–02 and 03).

Under fault conditions the voltage appearing at the earth electrode will radiate away from the electrode like the ripples radiating away from a pebble thrown into a pond. The voltage will fall to a safe level in the first two or three metres away from the point of the earth electrode as shown in Figure 8.26.

The basic method of measuring earth electrode resistance is to pass a current into the soil through the electrode and to measure the voltage required to produce this current.

Regulation 713–11 demands that where earth electrodes are used they should be tested. To make the test, either a hand operated tester or a mains energised double wound transformer with a separate ammeter and high resistance voltmeter is used. The test procedure is the same in both cases. The earth electrode is disconnected from all sources of the supply. An alternating current supplied by a double wound transformer (as shown in Figure 8.27) is passed between the earth electrode under test T and an auxiliary earth electrode T_1. The auxiliary electrode T_1 is placed at some distance from T so that the resistance area of the two electrodes do not overlap. A second auxiliary electrode T_2 is driven into the ground half way between T and T_1 and the voltmeter reading V tabulated. The resistance of the earth electrode is the voltmeter reading V, divided by the current flowing in the circuit and indicated on ammeter A.

To check that the resistance of the earth electrode is a true value, two further readings are taken at X and Y, with the auxiliary electrode T_2 moved 6 m further from and then 6 m nearer to T respectively. If the readings are substantially in agreement, the mean of the three readings is taken as the resistance of the earth electrode. If there is no agreement the test must be repeated with the distance between T and T_1 increased.

The test procedure is the same if a hand operated tester is used. The instrument is connected as shown in Figure 8.28. The hand operated generator is turned and the three dials rotated until null balance is indicated on the galvonometer. The value indicated by the dials

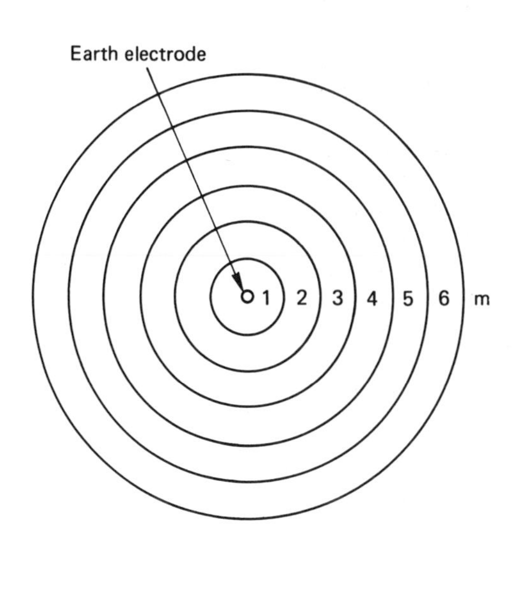

Figure 8.26 Earth electrode voltage gradient

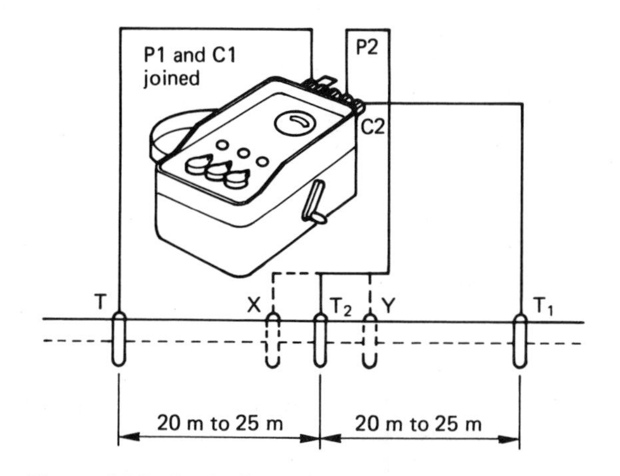

Figure 8.28 Earth electrode resistance test using a hand operated tester

gives the resistance of the earth electrode. Three readings are taken as in the previously described procedure and the average reading taken as the resistance of the earth electrode. The resistance of the earth electrode will depend upon the type of ground in which the electrode is driven. Wet, marshy land will give a lower resistance reading than rocky ground. Typical resistance readings are:

Marshy ground 5 Ω to 20 Ω
Agricultural soil 5 Ω to 50 Ω
Loam and clay 10 Ω to 150 Ω
Sandy gravel 200 Ω to 500 Ω
Rocky ground 500 Ω to 10 k Ω

Acceptable values of earth electrode resistance will be determined by the purpose for which the earth electrode is being used.

Lightning conductors provide a path of low resistance to lightning current, which may be many thousands of amperes. If the earth electrode forms the final connection for a lighting conductor it must have an electrode resistance of 10 Ω maximum (BS 6651 *The protection of structure against lightning*). The lightning protective system must be connected to the main earthing terminal of the electrical installation (Regulation 413–02–02).

In order that any protective device can operate under earth fault conditions it is necessary for an earth path to exist which can carry the fault current back to the supply transformer. In most installations this earth path is provided by the

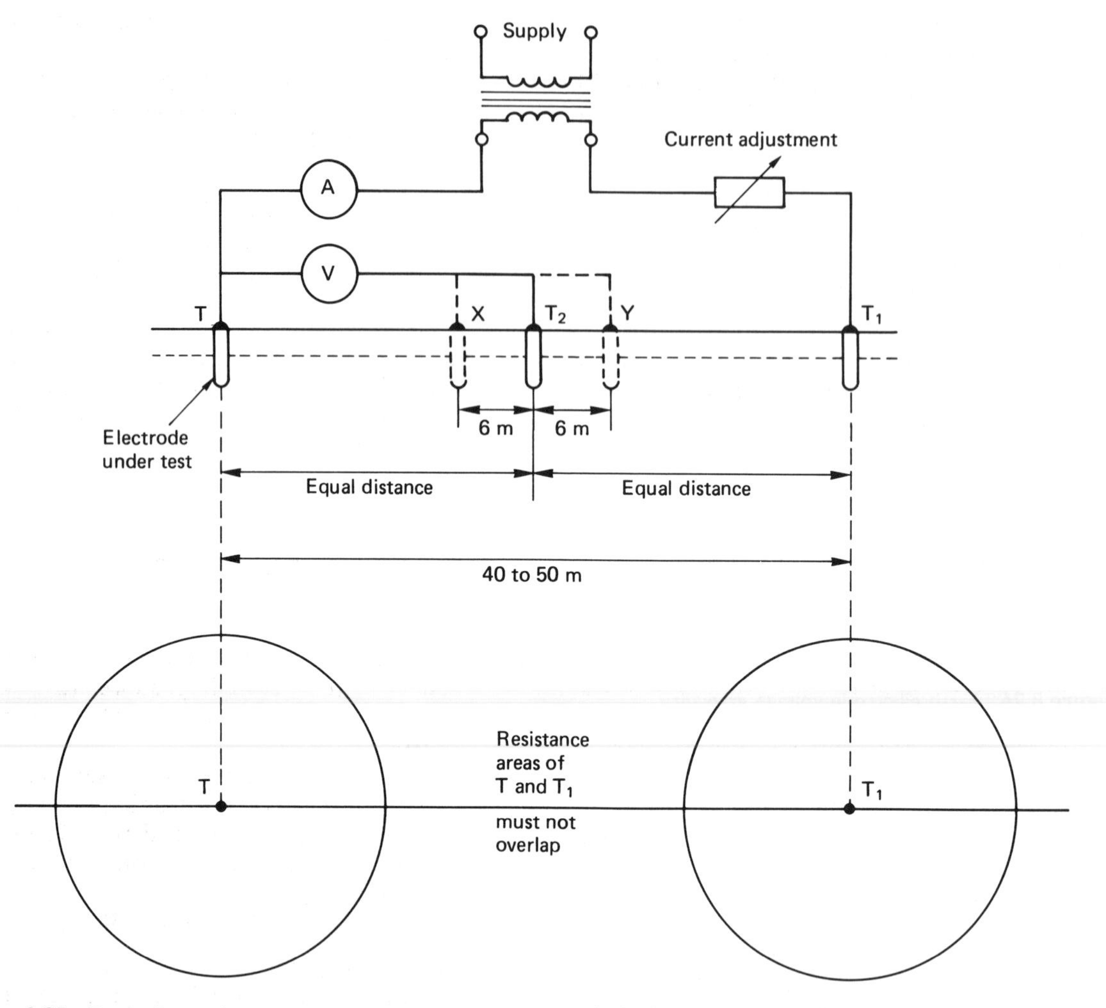

Figure 8.27 Earth electrode resistance test using a mains energised, double wound transformer

sheath of the supply cable, but in some rural areas where supplies are provided by overhead cables, the metallic sheath is not available and the general mass of earth is relied upon to provide the return path. The total resistance of the whole earth loop path must be low enough to permit the protective device to disconnect the supply to the circuit in 0.4 seconds for socket outlets and 5 seconds for fixed appliances (Regulations 413–02–09 and 13). The resistance of the earth electrode will probably be the biggest individual factor in the total resistance of the consumer's earth path.

Example 1

The total resistance of the complete earth path of an electrical installation supplied by a TT system is 20 Ω including the resistance of the consumers earth electrode. Calculate the earth fault current which would flow if the supply voltage was 240 V.

$$I = \frac{V}{R} \text{ (A)}$$

$$I = \frac{240 \text{ V}}{20 \, \Omega} = 12\text{A}$$

Under earth fault conditions only 12 A will flow which would not be sufficient to operate, for example, a 30 A ring main fuse, but would be sufficient to kill someone since 50 mA can be fatal. To operate a 30 A protective device effectively would require an earth electrode resistance of about 0.5 Ω. For this reason Regulation 471–08–06 recommends that socket outlets on a TT system be protected by RCCB. Regulation 413–02–16 states that the product of the earth loop impedance and the operating current of the RCCB should be less than 50.

If the electrode under test forms part of the earth return for a TT installation in conjunction with a residual current device, section 10.3.5 of the On Site Guide describes the following method:

1. disconnect the installation equipotential bonding from the earth electrode to ensure that the test current passes only through the earth electrode.
2. switch off the consumers unit to isolate the installation.
3. using a phase earth loop impedance tester, test between the incoming phase conductor and the earth electrode.

Record the result.

Section 4.2 of the On Site Guide tells us that the recommended maximum value of the earth fault loop impedance for a TT installation is 220 Ω.

Since most of the circuit impedance will be made up of the earth electrode resistance we can say that as acceptable value for the measurement of the earth electrode resistance would be less than about 200 Ω

Exercises

1 A meter with a moving coil movement:
(a) has a digital readout, (b) can be used on both a.c. and d.c. supplies, (c) has a linear scale, (d) can be used to measure power.

2 A meter with a moving iron movement:
(a) has a digital readout, (b) can be used on both a.c. and d.c. supplies, (c) has a linear scale, (d) can be used to measure power.

3 A dynamometer instrument:
(a) has a digital readout, (b) can only be used on electronic circuits, (c) has a linear scale, (d) can be used to measure power.

4 Damping in a moving coil instrument is achieved by:
(a) air vane, (b) air piston, (c) eddy currents, (d) spiral hair springs.

5 Instrument transformers can be used to extend the range of instruments connected to:
(a) a.c. circuits, (b) d.c. circuits, (c) 415 V supplies only, (d) rectified a.c. circuits.

6 A tong test instrument can also be correctly called:
(a) a dynamometer wattmeter (b) an insulation resistance tester (c) an earth loop impedance tester, (d) a clip-on ammeter.

7 To reduce errors when testing electronic circuits, the test instrument should:
(a) have a very low impedance, (b) have a very high impedance, (c) have a resistance equal to the circuit impedance, (d) have a resistance approximately equal to the circuit current.

8 The two wattmeter method is used to measure the power in a three phase three wire system. The two readings obtained were 100 W and 50 W and, therefore, the total power in the system is:
(a) 50 W, (b) 75 W, (c) 150 W, (d) 5 kW.

9 Before an ammeter can be removed from the secondary terminals of a current transformer connected to a load, the transformer terminals must be:
(a) open circuited, (b) short circuited, (c) connected to the primary winding, (d) connected to earth.

10 An acceptable earth electrode resistance test on a lightning conductor earth electrode must reveal a maximum value of:
(a) 10 Ω, (b) 100 Ω, (c) 0.5 MΩ, (d) 1 MΩ.

11 Use sketches to describe the operation of the deflection system in a moving coil instrument.

12 Explain how the basic moving coil system is modified so that a test instrument can be used on a.c. circuits.

13 Describe what is meant by damping of a system. Sketch a graph to show an over-damped, under-damped and critically damped system.

14 Describe eddy current damping

15 Describe air vane and air piston damping

16 A moving coil deflection system has a resistance of 5 Ω and gives full scale deflection when 15 mA flows through the moving coil. Calculate the value of the resistor required to make the movement into
(a) a 10 A ammeter, (b) a 250 V voltmeter. (c) Draw a circuit diagram showing how the resistor would be connected to the instrument movement in both cases.

17 Describe the construction and operation of a dynamometer wattmeter

18 Explain with the aid of a diagram how a single wattmeter can be used to measure the total power in
(a) a balanced three phase load, (b) an unbalanced three phase load.

19 Describe the construction and use of a voltage transformer

20 Describe the construction and use of a bar primary current transformer.

21 Draw a circuit diagram to show an ammeter and current transformer connected to measure the current in a single phase a.c. circuit. Explain why the secondary winding must not be open circuited when the transformer is connected to the supply.

22 Draw a circuit diagram showing a voltmeter and voltage transformer connected to measure the voltage in a single phase load.

23 Use a labelled sketch to describe the construction and operation of an energy meter.

24 Use a labelled sketch to describe the construction and operation of a tong tester.

25 Describe the construction and use of a phase sequence tester.

CHAPTER 9

Lighting

In ancient times, much of the indoor work done by man depended upon daylight being available to light the interior. Today almost all buildings have electric lighting installed and we automatically assume that we can work indoors or out of doors at any time of the day or night, and that light will always be available.

Good lighting is important in all building interiors, helping work to be done efficiently and safely and also playing an important part in creating pleasant and comfortable surroundings.

Lighting schemes are designed using many different types of light fitting or luminaire. Luminaire is the modern term given to the equipment which supports and surrounds the lamp and may control the distribution of the light. Modern lamps use the very latest technology to provide illumination cheaply and efficiently. To begin to understand the lamps and lighting technology used today, we must first define some of the terms we will be using.

Common lighting terms

Luminous intensity – symbol I

This is the illuminating power of the light source to radiate luminous flux in a particular direction. The earliest term used for the unit of luminous intensity was the candle power because the early standard was the wax candle. The SI unit is the candela (pronounced candeela) and is the unit of intensity of a point source which emits light energy from the source in all directions.

Luminous flux – symbol F

This is the flow of light which is radiated from a source. The SI unit is the lumen, one lumen being the light flux which is emitted within a unit solid angle (volume of a cone) from a point source of one candela.

Illuminance – symbol E

This is a measure of the light falling on a surface, which is also called the incident radiation. The SI unit is the Lux and is the illumination produced by one lumen over an area of one square metre.

Luminance – symbol L

Since this is a measure of the brightness of a surface it is also a measure of the light which is reflected from a surface. The objects we see vary in appearance according to the light which they emit or reflect toward the eye.

The SI units of luminance vary with the type of surface being considered. For a diffusing surface such as blotting paper or a matt white painted surface the unit of luminance is the lumen per square metre or Apostilb. With polished surfaces such as silvered glass reflector, the brightness is specified in terms of the light intensity and the unit is the candela per square metre.

Illumination laws

Rays of light falling upon a surface from some distance d will illuminate that surface with an illuminance of say one unit of Lux. If the distance d is doubled as shown in Figure 9.1, the one unit of Lux will fall over four square units of area. Thus the illumination of a surface follows the *inverse square law* where

$$E = \frac{I}{d^2} \text{ (lx)}$$

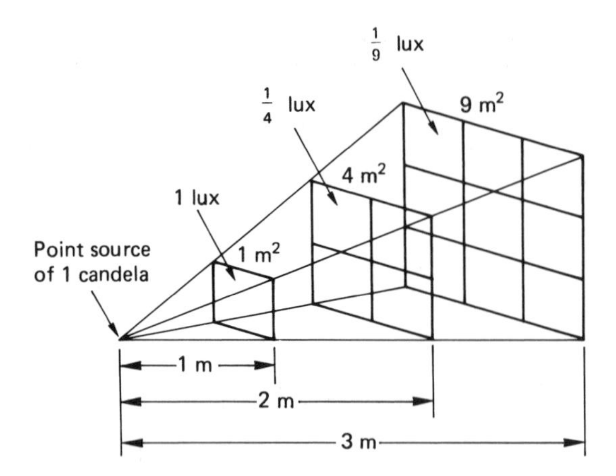

Figure 9.1 The inverse square law

Example 1
A lamp of luminous intensity 1000 candela is suspended 2 metres above a laboratory bench. Calculate the illuminance directly below the lamp

$$E = \frac{I}{d^2} \text{ (lx)}$$

$$\therefore \quad E = \frac{1000 \text{ cd}}{2 \times 2 \text{ m}} = 250 \text{ lx.}$$

The illumination of surface A in Figure 9.2 will follow the inverse square law described above. If this surface were removed, the same luminous flux would then fall on surface B. Since the parallel rays of light falling on the inclined surface B are spread over a larger surface area, the illuminance will be reduced by a factor θ and therefore

$$E = \frac{I \cos \theta}{d^2} \text{ (lx).}$$

Since the two surfaces are joined together by the trigonometry of the cosine rules this equation is known as the *cosine law*.

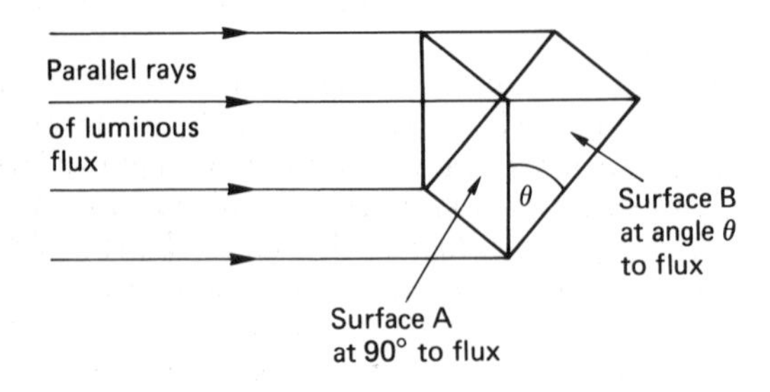

Figure 9.2 The cosine law

Example 2
A street lantern suspends a 2000 cd light source 4 m above the ground. Determine the illuminance directly below the lamp and 3 m to one side of the lamp base.

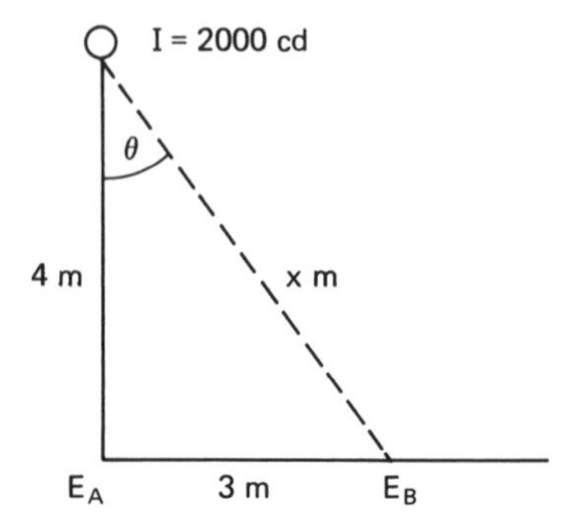

Illuminance below the lamp E_A

$$E_A = \frac{I}{d^2} \text{ (lx)}$$

$$\therefore \quad E_A = \frac{2000 \text{ cd}}{(4 \text{ m})^2} = 125 \text{ lx.}$$

Illuminance at 3 m to one side of the lantern E_B: The distance between the light source and the position on the ground at E_B can be found by Pythagoras Laws.

$$x \text{ (m)} = \sqrt{(4 \text{ m})^2 + (3 \text{ m})^2} = \sqrt{25} \text{ m}$$

$$x = 5 \text{ m}$$

$$E_B = \frac{I \cos \theta}{d^2} \text{ (lx) and } \cos \theta = \frac{4}{5}$$

$$\therefore \quad E_B = \frac{2000 \text{ cd} \times 4}{(5 \text{ m})^2 \times 5} = 64 \text{ lx.}$$

Example 3
A discharge lamp is suspended from a ceiling 4 m above a bench. The illuminance on the bench below the lamp was 300 Lux. Find:
(a) the luminous intensity of the lamp,
(b) the distance along the bench where the illuminance falls to 153.6 Lux.

For (a), $\quad E_A = \frac{I}{d^2} \text{ (lx)} \quad \therefore \quad I = E_A\, d^2 \text{ (cd)}$

$$I = 300 \text{ lx} \times 16 \text{ m} = 4800 \text{ cd.}$$

For (b), $\quad E_B = \frac{I}{d^2} \cos \theta \text{ (lx)}$

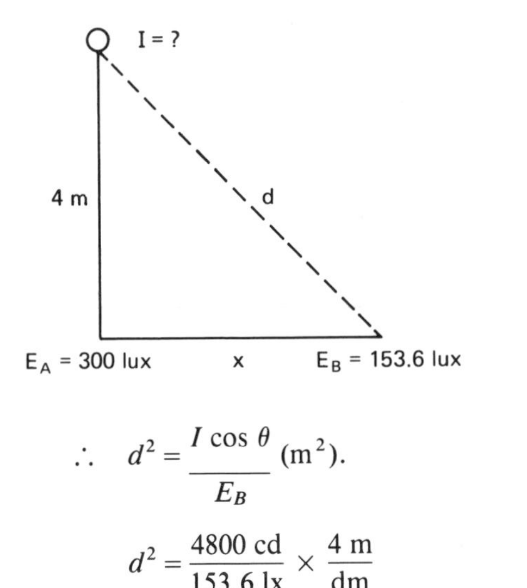

$$\therefore \quad d^2 = \frac{I \cos \theta}{E_B} \ (\text{m}^2).$$

$$d^2 = \frac{4800 \text{ cd}}{153.6 \text{ lx}} \times \frac{4 \text{ m}}{d\text{m}}$$

$$d^3 = 125 \qquad \therefore \quad d = \sqrt[3]{125} = 5 \text{ m}.$$

By Pythagoras, $x = \sqrt{5^2 - 4^2} = 3$ m

Measurement of illuminance

To take a reading place a suitable illuminance meter, calibrated in Lux, onto the surface whose illumination level is to be measured. For general light levels hold the instrument 85 cm above the floor in a horizontal plane.

Take readings from the appropriate scale but take care not to obscure the photocell in any way when taking measurements, for example, by casting a shadow with the body or the hand. The recommended levels of illuminance for various types of installation are given by the IES code which is usually printed on the back of the meter, as shown in Figure 9.3. Some examples are given in Table 9.1.

The activities being carried out in a room will determine the levels of illuminance required since different levels of illumination are required for the successful operation or completion of different tasks. The assembly of electronic components in a factory will require a higher level of illumination than, say, the assembly of engine components in a garage because the electronic components are much smaller and finer detail is required for their successful assembly.

The inverse square law calculations considered earlier are only suitable for designing lighting schemes where there are no reflecting surfaces producing secondary additional illumination. This method could be used to design an outdoor lighting scheme for a cathedral, bridge or public building.

Interior luminaires produce light directly on to the working surface but additionally there is a secondary source of illumination from light reflected from the walls and ceilings. When designing interior lighting schemes the method most frequently used depends upon a determination of the total flux required to provide a given value of illuminance at the working plane. This method is generally known as the Lumen Method.

Table 9.1 Illuminance values

Task	Working Situation	Illuminance (Lux)
Casual seeing	Storage rooms, stairs and washrooms	100
Rough assembly	Workshops and garages	300
Reading, writing and drawing	Classrooms and offices	500
Fine assembly	Electronic component assembly	1000
Minute assembly	Watchmaking	3000

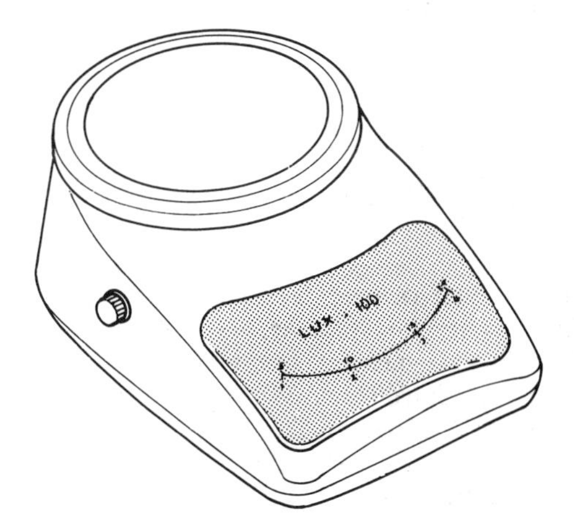

Figure 9.3 A typical lightmeter

The Lumen Method

To determine the total number of luminaires required to produce a given illuminance by the Lumen Method we apply the following formulae:

$$\text{Total number of luminaires required to provide a chosen level of illumination at a surface} = \frac{\text{illuminance level (lx)} \times \text{area (m}^2\text{)}}{\text{lumen output of each luminaire (lm)} \times \text{UF} \times \text{MF}}$$

where

- the illuminance level is chosen after consideration of the IES code,
- the area is the working area to be illuminated,
- the lumen output of each luminaire is that given in the manufacturers specification and may be found by reference to tables such as that given in Table 9.2,
- UF is the utilisation factor,
- MF is the maintenance factor.

Table 9.2 Characteristics of a Thorn Lighting 1500 mm 65 watt Bi-Pin tube

Tube colour	Lighting design lumens*	Colour rendering quality	Colour appearance
Artificial daylight	2100	Excellent	Cool
De Luxe Natural	2500	Very good	Intermediate
De Luxe Warm white	3200	Good	Warm
Natural	3400	Good	Intermediate
Daylight	4450	Fair	Cool
Warm white	4600	Fair	Warm
White	4750	Fair	Warm
Red	250	Poor	Deep red

* The lighting design lumens are the output lumens after 2000 hours.

Burning position	Lamp may be operated in any position
Rated life	7500 hours
Efficacy	30 to 70 lm/W depending upon the tube colour

Utilisation factor

The light flux reaching the working plane is always less than the lumen output of the lamp since some of the light flux is absorbed by the luminaire and some is directed to walls, ceilings and floors where only a small part is reflected. The utilisation factor is given by

$$\text{UF} = \frac{\text{Total light reaching the working surface}}{\text{Total light flux emitted by the lamp}}.$$

The UF is expressed as a number which is always less than unity and a typical value might be 0.9 for a modern office building.

Maintenance factor

The light output of a luminaire is reduced during its life because of an accumulation of dust and dirt on the lamp and fitting. Decorations also deteriorate with time, and this results in more light flux being absorbed by the walls and ceiling. A figure of about 0.8 is normally taken to account for this loss of light to the surroundings but in very dusty, dirty or smoky atmospheres the number may be further reduced.

Example

It is proposed to illuminate an electronic workshop, $9\ \text{m} \times 8\ \text{m} \times 3\ \text{m}$ high to an illuminance of 500 Lux at the bench level. The specification calls for luminaires having one 1500 mm 65 W natural tube having an output of 3400 lumens (see Table 9.2). Determine the number of luminaires required for this installation when the UF and MF are 0.9 and 0.8 respectively.

$$\text{The number of luminaires required} = \frac{E\ (\text{lx}) \times \text{area (m}^2\text{)}}{\text{lumens from each luminaire} \times \text{UF} \times \text{MF}}$$

$$\text{Number of luminaires} = \frac{500\ \text{lx} \times 9\ \text{m} \times 8\ \text{m}}{3400 \times 0.9 \times 0.8} = 14.7.$$

Therefore 15 luminaires will be required.

Space to mounting height ratio

The correct mounting height of luminaires is important since glare may result if fittings are placed in the line of vision. Excessive height will result in a rapid reduction of illuminance, as demonstrated by the inverse square law, and make lamp replacement and maintenance difficult. The correct spacing of luminaires is important since large spaces between the fittings may result in a fall-off of illuminance at the working plane midway between adjacent fittings. The illuminance between the luminaires must not be allowed to fall below 70% of the value directly below the fitting. For most installations a spacing to mounting height ratio of 1 : 1 to 2 : 1 above the working surface is usually considered adequate and the working surface is normally taken as 0.85 m above the floor level as shown in Figure 9.4.

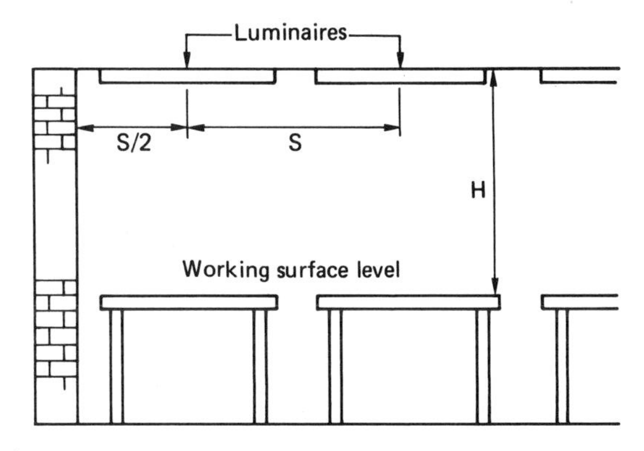

Figure 9.4 Space to height ratio (S : H)

Layout of luminaires

To maintain an even distribution of illuminance from the luminaires, those adjacent to the walls of the room should be fixed at half the spacing distance. This is because a point in the middle of the room receives luminous flux from two adjacent luminaires, whilst a point close to the wall is illuminated mainly from only one luminaire.

Considering the previous example of an electronic workshop requiring 15 luminaires to provide the required illuminance, if we assume a space to height ratio of 1 : 1, the best layout may be four rows of four luminaires to each row. This would necessitate using one extra luminaire than the calculations suggested. This is quite acceptable since the overall illuminance will be raised by only about 6% and the resultant layout will be more symetrical whilst complying very closely with the space to height ratio. The layout is shown in Figure 9.5.

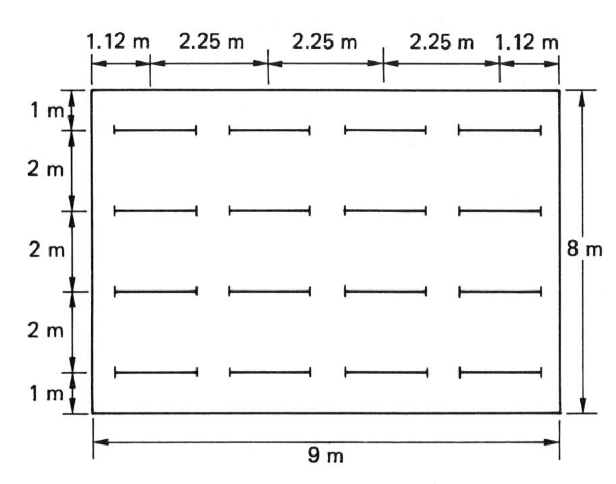

The mounting height in this case is the ceiling height minus the height of the working surface ∴ H = 3.0 – 0.85 = 2.15 m

Figure 9.5 Layout of luminaires

Comparison of light sources

When comparing one light source with another we are interested in the colour reproducing qualities of the lamp and the efficiency with which the lamp converts electricity into illumination. These qualities are expressed by the lamp's colour rendering qualities and its efficacy.

Lamp efficacy

The performance of a lamp is quoted in terms of the number of lumens of light flux which it emits to the electrical energy input which it consumes. Thus efficacy is measured in lumens per watt; the greater the efficacy the better is the lamps performance in converting electrical energy into light energy.

Colour rendering

We recognise various materials and surfaces as having a particular colour because luminous flux of a frequency corresponding to that colour is reflected from the surface to our eye which is then processed by our brain. White light is made up of the combined frequencies of the colours red, orange, yellow, green, blue, indigo and violet. Colours can only be seen if the lamp supplying the illuminance is emitting light of that particular frequency. The ability to show colours faithfully is a measure of the colour rendering property of the light source.

GLS lamps

General lighting service lamps produce light as a result of the heating effect of an electrical current. A fine tungsten wire is first coiled and coiled again to form the incandescent filament of the GLS lamp. The coiled coil arrangement reduces filament cooling and increases the light output by allowing the filament to operate at a higher temperature. The light output covers the visible spectrum giving a warm white to yellow light with a colour rendering quality classified as fairly good. The efficacy of the GLS lamp is 14 lumens per watt over its intended lifespan of 1000 hours.

The filament lamp in its simplest form is a purely functional light source which is unchallenged on the domestic market despite the manufacture of more efficient lamps. One factor which may have contributed to its popularity is that lamp designers have been able to modify the glass envelope of the lamp to give a very pleasing decorative appearance as shown by Figure 9.6.

Tungsten halogen lamps

In the GLS lamp, the high operating temperature of the filament causes some evaporation of the tungsten which is carried by convection currents on to the bulb wall. When the lamp has been in service for some time, evaporated tungsten darkens the bulb wall with the result that the light output is reduced and the filament becomes thinner and eventually fails.

To overcome these problems the envelope of the tungsten halogen lamp contains a trace of one of the halogen gasses; iodine, chlorine, bromine or fluorine. This allows a reversible chemical reaction to occur between the tungsten filament and the halogen gas. When tungsten is evaporated from the incandescent filament, some part of it spreads out towards the bulb wall, but at a point close to the wall where the temperature conditions are favourable, the tungsten combines with the halogen. This tungsten halide molecule then drifts back toward the filament where it once more separates depositing the tungsten back on to the filament, leaving the halogen available for a further reaction cycle.

Since all the evaporated tungsten is returned to the filament, the bulb blackening normally

Figure 9.6 Some decorative GLS lamp shapes

Lamp characteristics

Watts	Lighting design lumens
40	380 at 240 V
60	660 at 240 V
100	660 at 240 V
150	2000 at 240 V

Burning position	Lamp may be operated in any position
Rated life	1000 hours
Efficacy	14 lm/W
Colour rendering	Fairly good

associated with tungsten lamps is completely eliminated and a high efficacy is maintained throughout the life of the lamp.

A minimum bulb wall temperature of 250°C is required to maintain the halogen cycle and consequently a small glass envelope is required. This also permits a much higher gas pressure to be used which increases the lamp life to 2000 hours and allows the filament to be operated at a higher temperature, giving more light. The lamp is very small and produces a very white intense light giving it a colour rendering classification of good and an efficacy of 20 lumens per watt.

The research and development of the tungsten halogen lamp, as shown in Figure 9.7, was a major development in lamp design and resulted in Thorn Lighting gaining the Queen's Award for Technical innovation in Industry in 1972.

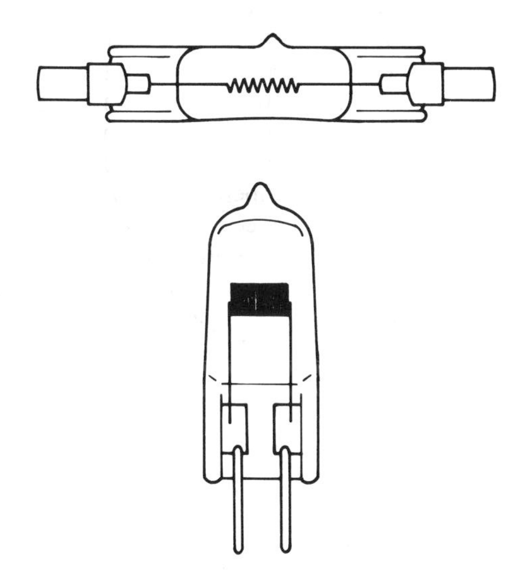

Lamp characteristics

Watts	Lighting design lumens
300	5000 at 230 V
500	9500 at 230 V

Burning position	Linear lamps must be operated horizontally or within 4° of the horizontal
Rated life	2000 hours
Efficacy	20 lm/W
Colour rendering	Good

When installing the lamp, grease contamination of the glass envelope by touching must be avoided. Any grease present on the outer surface will cause cracking and premature failure of the lamp because of the high operating temperatures. A paper sleeve should be used when handling the lamp, and if accidentally touched with bare hands the lamp should be cleaned with methylated spirits.

Figure 9.7 A tungsten halogen lamp

Discharge lamps

Discharge lamps do not produce light by means of an incandescent filament but by the excitation of a gas or metallic vapour contained within a glass envelope. A voltage applied to two terminals or electrodes sealed into the end of a glass tube containing a gas or metallic vapour will excite the contents and produce light directly.

The colour of the light produced depends upon the type of gas or metallic vapour contained within the tube. Some examples are given below:

Gases	neon	red
	argon	green/blue
	hydrogen	pink
	helium	ivory
	mercury	blue
Metallic vapours	sodium	yellow
	magnesium	grass green

Let us now consider four of the more frequently used discharge lamps.

Fluorescent tube

A fluorescent lamp is a linear arc tube, internally coated with a fluorescent powder, containing a low pressure mercury vapour discharge and given the designation MCF by lamp manufacturers. The lamp construction is shown in Figure 9.8 and the characteristics of the variously coloured tubes are given in Table 9.2.

Passing a current through the cathodes of the tube causes them to become hot and produce a cloud of electrons which ionise the gas in the immediate vicinity of the cathodes. This ionisation then spreads to the whole length of the tube producing invisible ultra-violet rays and some blue light. The fluorescent powder on the inside of the tube is sensitive to ultra-violet rays and converts

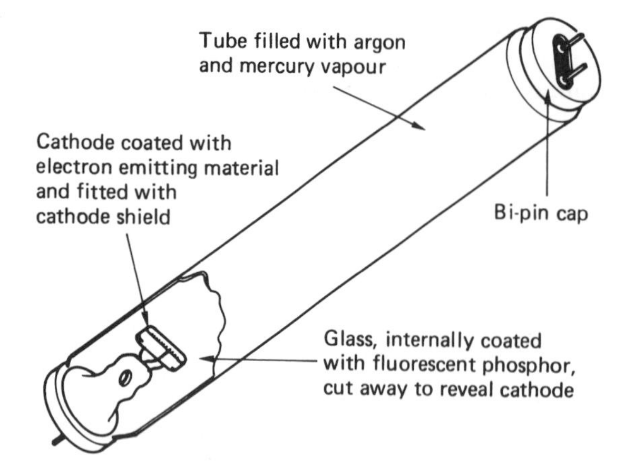

Figure 9.8 Fluorescent lamp construction

this radiation into visible light. The fluorescent powder on the inside of the tube can be mixed to give light of almost any desired colour or grade of white light. Some mixes have their maximum light output in the yellow-green region of the spectrum giving maximum efficacy but poor colour rendering. Other mixes give better colour rendering at the cost of reduced lumen output as can be seen from Table 9.2. The lamp has many domestic, industrial and commercial applications.

High pressure mercury vapour lamp

The high pressure mercury discharge takes place in a quartz glass arc tube contained within an outer bulb which, in the case of the lamp classified as MBF, is internally coated with fluorescent powder. The lamp's construction and characteristics are shown in Figure 9.9.

The inner discharge tube contains the mercury vapour and a small amount of argon gas to assist starting. The main electrodes are positioned at either end of the tube and a starting electrode is positioned close to one main electrode.

When the supply is switched on the current is insufficient to initiate a discharge between the main electrodes, but ionisation does occur between the starting electrode and one main electrode in the argon gas. This spreads through the arc tube to the other main electrode. As the lamp warms the mercury is vapourised, the pressure builds up and the lamp achieves full brilliance after about five to seven minutes.

If the supply is switched off the lamp cannot be re-lit until the pressure in the arc tube has reduced. It may take a further five minutes to re-strike the lamp.

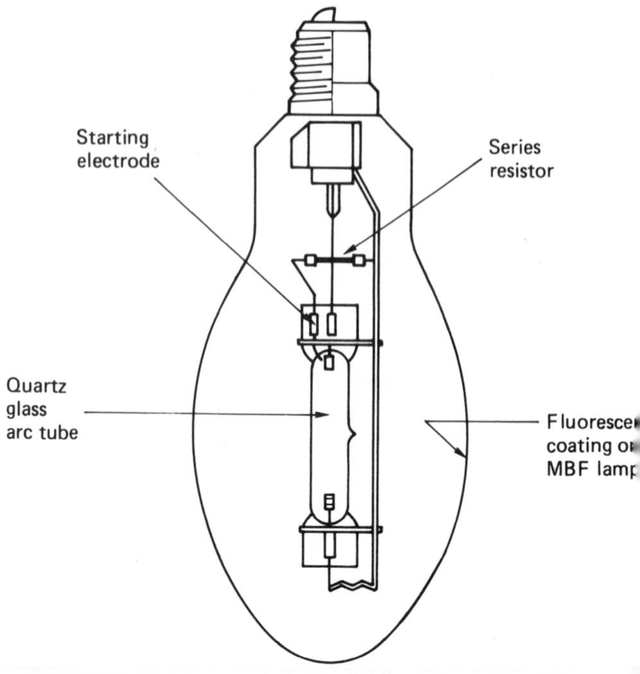

Lamp Characteristics

Watts	Lighting design lumens
50	1800
80	3350
125	5550
250	12000
400	21500
700	38000
1000	54000

Burning position	Lamp may be operated in any position
Rated life	7500 hours
Efficacy	38 to 56 lm/W
Colour rendering	Fairly good

Figure 9.9 High pressure mercury vapour lamp

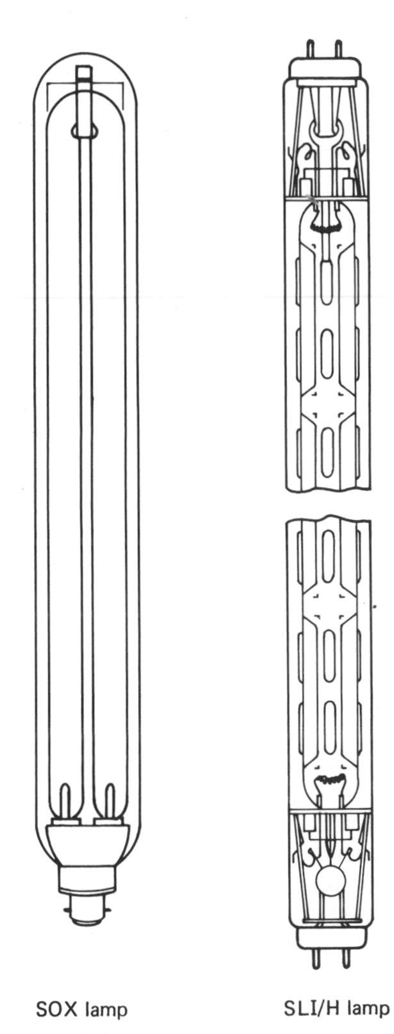

Lamp characteristics

Watts	Lighting design (lumens
Type SOX	
35	4300
55	7500
90	12500
135	21500
Type SLI/H	
140	20000
200	25000
200 HO	27500

Burning position	Horizontal or within 20° of the horizontal
Rated life	6000 hours
Guaranteed life	4000 hours
Efficacy	61 to 160 lm/W
Colour rendering	Very poor

Figure 9.10 Low pressure sodium lamp

The lamp is used for commercial and industrial installations, street lighting, shopping centre illumination and area floodlighting.

Low pressure sodium lamps

The low pressure sodium discharge takes place in a U-shaped arc tube made of special glass which is resistant to sodium attack. This U-tube is encased in a tubular outer bulb of clear glass as shown in Figure 9.10. Lamps classified as type SOX have a BC lampholder while the SLI/H lamp has a bi-pin lampholder at each end.

Since at room temperature the pressure of sodium is very low, a discharge cannot be initiated in sodium vapour alone. Therefore, the arc tube also contains neon gas to start the lamp. The arc path of the low pressure sodium lamp is much longer than that of mercury lamps and starting is achieved by imposing a high voltage equal to about twice the main voltage across the electrodes by means of a leakage transformer. This voltage initiates a discharge in the neon gas which heats up the sodium. The sodium vaporises and over a period of six to eleven minutes the lamp reaches full brilliance, changing colour from red to bright yellow.

The lamp must be operated horizontally so that when the lamp is switched off the condensing sodium is evenly distributed around the U-tube.

The light output is yellow and has poor colour rendering properties but this is compensated by the fact that the wavelength of the light is close to that at which the human eye has its maximum sensitivity, giving the lamp a high efficacy. The main application for this lamp is street lighting where the light output meets the requirements of the Ministry of Transport.

High pressure sodium lamp

The high pressure sodium discharge takes place in a sintered aluminium oxide arc tube contained within a hard glass outer bulb. Until recently no suitable material was available which would withstand the extreme chemical activity of sodium at high pressure. The construction and characteristics of the high pressure sodium lamp classified as type SON are given in Figure 9.11.

The arc tube contains sodium and a small amount of argon or xenon to assist starting. When the lamp is switched on an electronic pulse ignitor

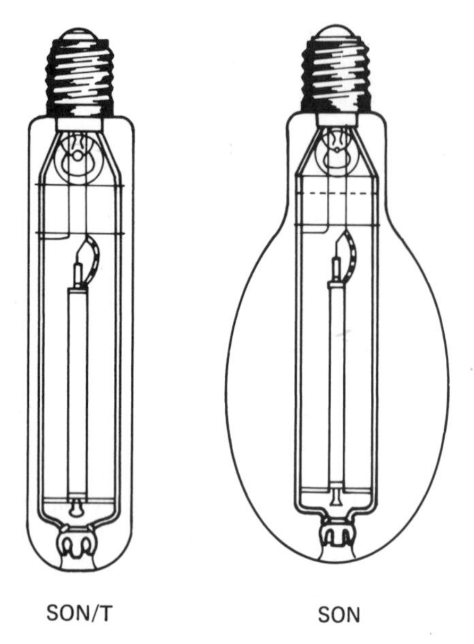

Lamp characteristics

Watts	Lighting design lumens
Tubular clear (SON/T)	
250	21 000
400	38 000
Elliptical coated (SON)	
250	19 500
400	36 000

Burning position	Universal
Rated life	6000 hours
Guaranteed life	4000 hours
Efficacy	100 to 120 lm/W
Colour rendering	Fair

Figure 9.11 High pressure sodium lamp

of 2 kV or more initiates a discharge in the starter gas. This heats up the sodium and in about five to seven minutes the sodium vaporises and the lamp achieves full brilliance. Both colour and efficacy improve as the pressure of the sodium rises giving a pleasant golden white colour to the light which is classified as having a fair colour rendering quality.

The SON lamp is suitable for many applications. Because of the warming glow of the illuminance it is used in food halls and hotel reception areas. Also, because of the high efficacy and long lamp life it is used for high bay lighting in factories and warehouses and for area flood lighting at airports, car parks and dockyards.

Control gear for lamps

Luminaires are wired using the standard lighting circuits described in Chapter 8 of *Basic Electrical Installation Work*, but discharge lamps require additional control gear and circuitry for their efficient and safe operation. The circuit diagrams for high pressure mercury vapour and high and low pressure sodium lamps are given in Figure 9.12. Each of these circuits requires the inclusion of a choke or transformer creating a lagging power factor which must be corrected. This is usually achieved by connecting a capacitor across the supply to the luminaire as shown.

Fluorescent lamp control circuits

A fluorescent lamp requires some means of initiating the discharge in the tube, and a device to control the current once the arc is struck. Since the lamps are usually operated on a.c. supplies, these functions are usually achieved by means of a choke ballast. Three basic circuits are commonly used to achieve starting; switch-start, quick-start and semi-resonant.

Switch-start fluorescent lamp circuit

Figure 9.13 shows a switch-start fluorescent lamp circuit in which a glow type starter switch is now standard. A glow type starter switch consists of two bi-metallic strip electrodes encased in a glass bulb containing an inert gas. The starter switch is initially open circuit. When the supply is switched on the full mains voltage is developed across these contacts and a glow discharge takes place between them. This warms the switch electrodes and they bend toward each other until the switch makes contact. This allows current to flow through the lamp electrodes which become heated so that a cloud of electrons are formed at each end of the tube which glows.

When the contacts in the starter switch are made the glow discharge between the contacts is extinguished since no voltage is developed across the switch. The starter switch contacts cool and after a few seconds spring apart. Since there is a choke in

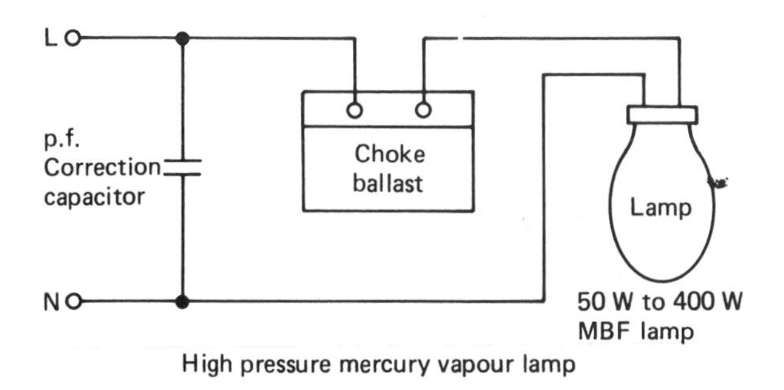

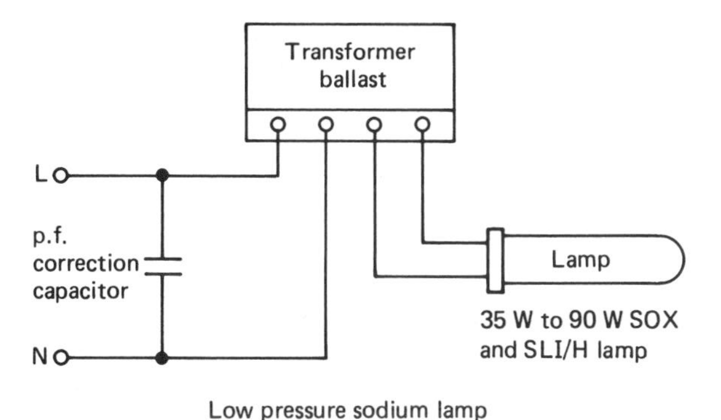

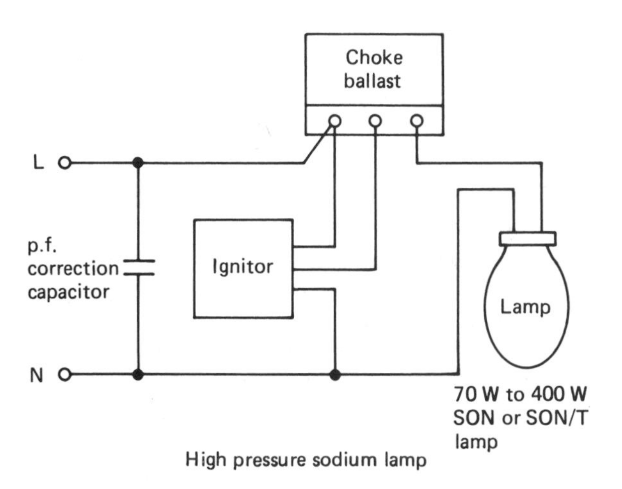

Figure 9.12 Discharge lamp control gear circuits

series with the lamp, the breaking of this inductive circuit causes a voltage surge across the lamp electrodes which is sufficient to strike the main arc in the tube. If the lamp does not strike first time the process is repeated.

When the main arc has been struck in the low pressure mercury vapour, the current is limited by

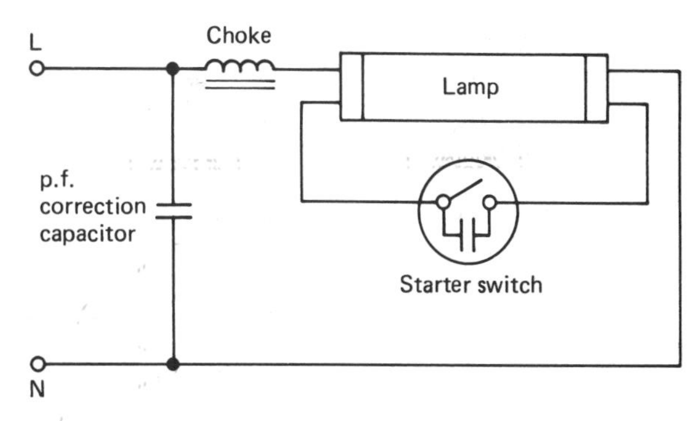

Figure 9.13 Switch-start fluorescent lamp circuit

the choke. The capacitor across the mains supply provides power factor correction and the capacitor across the starter switch contact is for radio interference supression.

Quick-start fluorescent lamp circuit

When the circuit is switched on the tube cathodes are heated by a small auto-transformer. After a short preheating period the mercury vapour is ionised and spreads rapidly through the tube to strike the arc. The luminaire or some earthed metal must be in close proximity to the lamp to assist in the striking of the main arc. In some cases a metal strip may be bonded along the tube length to assist starting.

When the main arc has been struck the current flowing in the circuit is limited by the choke. A capacitor connected across the supply provides p.f. correction. The circuit is shown in Figure 9.14.

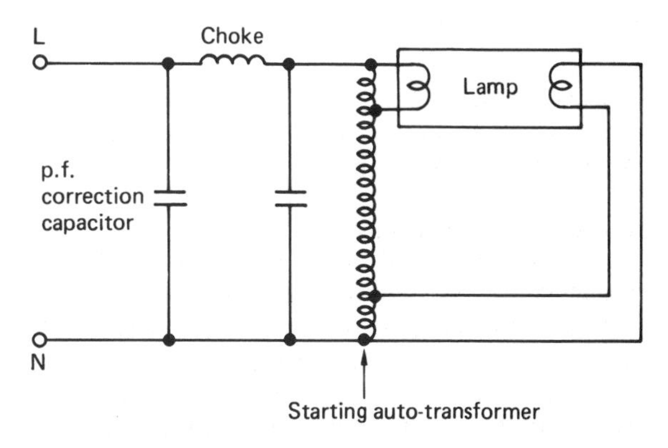

Figure 9.14 Quick-start fluorescent lamp circuit

Semi-resonant start fluorescent lamp circuit

In this circuit a specially wound transformer takes the place of the choke. When the circuit is switched on, a current flows through the primary winding to one cathode of the lamp, through the secondary winding and a large capacitor to the other cathode. The secondary winding is wound in opposition to the primary winding. Therefore, the voltage developed across the transformer windings is 180° out of phase.

The current flowing through the electrodes causes an electron cloud to form around each cathode. This cloud spreads rapidly through the tube due to the voltage across the tube being increased by winding the transformer windings in opposition. When the main arc has been struck the current is limited by the primary winding of the transformer which behaves as a choke. A p.f. correction capacitor is not necessary since the circuit is predominantly capacitive and has a high power factor. With the luminaire earthed to assist starting this circuit will start very easily at temperatures as low as −5°C. The circuit is shown in Figure 9.15.

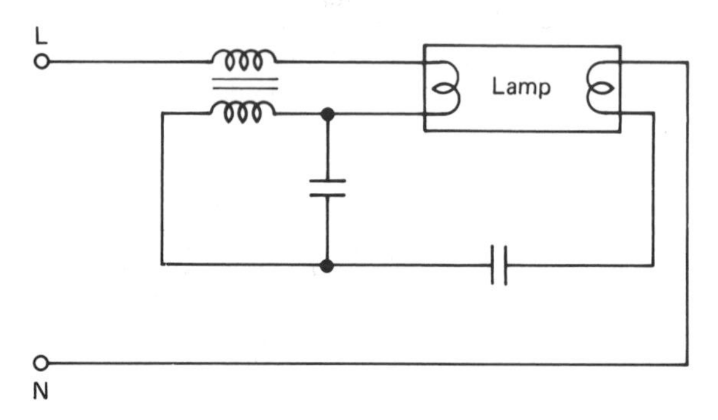

Figure 9.15 Semi-resonant start fluorescent lamp circuit

Installation of luminaires

Operating position

Some lamps, particularly discharge lamps, have limitations placed upon their operating position. Since the luminaire is designed to support the lamp, any restrictions upon the operating position of the lamp will affect the position of the luminaire. Some indications of the operating position of lamps were given earlier under the individual lamp characteristics. The luminaire must be suitable for the environment in which it is to operate. This may be a corrosive atmosphere, an outdoor situation or a low temperature zone. On the other hand, the luminaire may be required to look attractive in a commercial environment. It must satisfy all these requirements while at the same time providing adequate illumination without glare.

Many lamps contain a wire filament and delicate supports enclosed in a glass envelope. The luminaire is designed to give adequate support to the lamp under normal conditions, but a luminaire subjected to excessive vibration will encourage the lamp to fail prematurely either by breaking the filament, cracking the glass envelope or breaking the lampholder seal.

Control gear

Chokes and ballasts for discharge lamps have laminated sheet steel cores in which a constant reversal of the magnetic field due to the a.c. supply sets up vibrations. In most standard chokes the noise level is extremely low and those manufactured to BS 2818 have a maximum permitted noise level of 30 dB. This noise level is about equal to the sound produced by a loud pocket watch or 'Swatch' at a distance of one metre.

Chokes must be rigidly fixed, otherwise metal fittings can amplify choke noise. Plasterboard, hardboard or wooden panels can also act as a sounding board for control gear or luminaires mounted upon them, thereby amplifying choke noise. The background noise will obviously affect peoples' ability to detect choke noise, and so control gear and luminaires which would be considered noisy in a library or church may be unnoticeable in a busy shop or office.

The cable, accessories and fixing box must be suitable for the mass suspended and the ambient temperature in accordance with Regulations 553–03–03 and 554–01–01. Self-contained luminaires must have an adjacent means of isolation provided in addition to the functional switch to facilitate safe maintenance and repair (Regulations 476–02–04).

Control gear should be mounted as close as possible to the lamp. Where it is liable to cause overheating it must be either

- enclosed in a suitably designed non-combustible enclosure or
- mounted so as to allow heat to dissipate, or placed at a sufficient distance from adjacent materials to prevent the risk of fire (Regulations 422–01–01 and 02).

Discharge lighting may also cause a stroboscopic effect where rotating or reciprocating machinery is being used. This effect causes rotating machinery to appear stationary. The elimination of this dangerous effect is discussed in Chapter 8 of *Basic Electrical Installation Work*.

Loading and switching of discharge circuits

Discharge circuits must be capable of carrying the total steady current (the current required by the lamp plus the current required by any control gear). Appendix 1 of the On Site Guide states that where more exact information is not available, the rating of the final circuits for discharge lamps may be taken as the rated lamp watts multiplied by 1.8. Therefore, an 80 watt fluorescent lamp luminaire will have an assumed demand of $80 \times 1.8 = 144$ watts.

All discharge lighting circuits are inductive and will cause excessive wear of the functional switch contacts. Where discharge lighting circuits are to be switched, the rating of the functional switch must be suitable for breaking an inductive load.

High voltage discharge lighting

The popularity of high voltage discharge lighting has arisen almost entirely from its use in advertising. Piccadilly Circus in London attracts visitors from around the world to admire the high voltage discharge signs and the statue of Eros.

A gas sealed into a glass tube with electrodes at each end will ionise when a high voltage is applied to the electrodes. The ionised gas will glow with a colour which is characteristic of the gas contained in the tube. The luminous flux radiated by hydrogen is pink, argon green/blue and neon red. Since many advertisement signs incorporate neon gas, the term 'neon sign' has become synonymous with high voltage discharge signs.

The glass letters of the advertisement sign may be built up individually and joined together in series with nickel wire enclosed in a thin glass tube. Alternatively, the words may be formed from one piece of glass tube, the glass links between the letters being painted over so that only the letters illuminate. (See Figure 9.16).

Individual letters are formed by heating the glass tube with a blowtorch and bending it into the

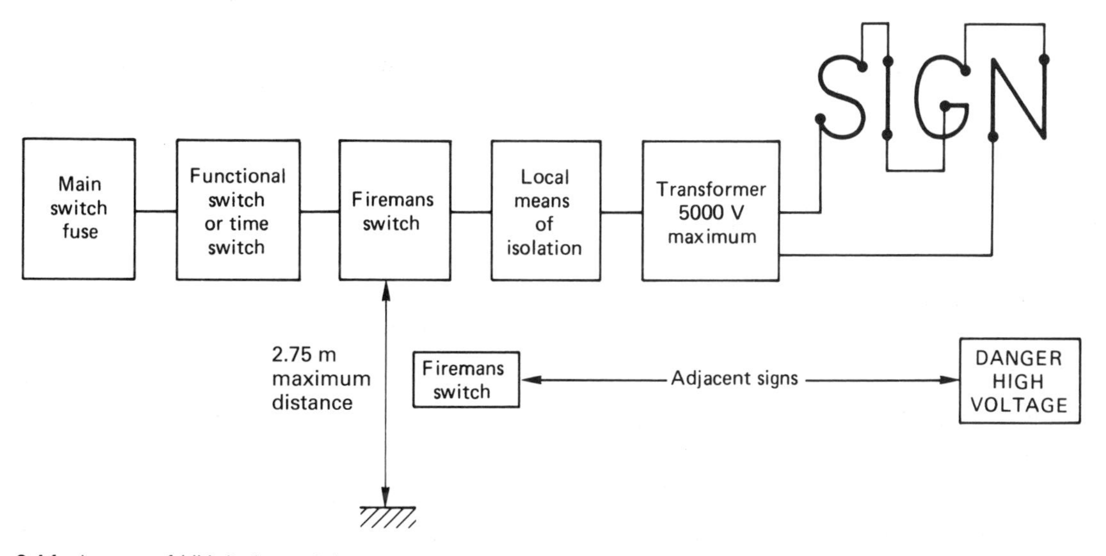

Figure 9.16 Layout of HV discharge lighting circuit

required shape. An electrode is then sealed to the end of the tube and injected with the desired gas.

The diameter of the tube will to some extent be influenced by the size of the letters to be formed, but common sizes are 10, 15, 20 and 30 mm. These tubes typically carry currents of 25, 35, 60 and 150 mA respectively.

$$\text{Total volt drop} = \text{Volt drop in tube} + \text{Volt drop across electrodes}$$

For neon gas,

V.D. per pair of electrodes = 300 V
V.D. per metre of 15 mm tube = 400 V

Example

A high voltage neon filled discharge sign spells the word 'Victoria' in separate letters made from 19 m of 15 mm glass tube. Calculate the secondary voltage of the transformer and the output power if the p.f. is 0.8.

Since the word 'Victoria' is made up of eight individual letters, eight pairs of electrodes will be required.

V.D. = V.D. in tube + V.D. across electrodes
V.D. = $(19 \text{ m} \times 400 \text{ V}) + (8 \times 300 \text{ V})$
V.D. = 7600 V + 2400 V = 10 kV

The sign could be fed from a 10 kV centre tapped transformer since this would give 5 kV to earth and satisfy the requirements of B.S. 559.

Power = $VI \cos\theta$ (W)
Power = $10\,000 \times 35 \times 10^{-3} \times 0.8 = 280$ W.

When installing high voltage discharge lighting special attention should be paid to Regulations 554–02. A means of isolation complying with Regulation 476–02–04 must be provided at the point of supply and also adjacent to the control equipment for safe maintenance. The means of isolation should preferably be provided with a lock or removable handle so that the supply cannot accidentally be switched on when maintenance is being carried out.

The secondary voltage of the transformer must not exceed 5 kV to earth. Any transformer with an output of more than 500 W must be protected by a circuit breaker. The auxillary equipment such as transformers and capacitors must be contained in substantial metal enclosures and a 'Danger–High Voltage' notice placed on each container.

A fireman's emergency switch must be provided adjacent to the discharge lamps for all exterior installations and for interior installations which run unattended. The switch must be red with the off position at the top and mounted not more than 2.75 m from the ground (Regulations 476–03–05 and 537–04–06).

On interior installations the switch should be installed in the main entrance and a warning sign should be fixed alongside, stating 'Fireman's Switch' in letters not less than 13 mm high.

For the purpose of these regulations a high voltage discharge sign installation installed in a closed market or arcade is considered to be an exterior installation. Only small portable discharge lighting, luminaires or signs with a rating not exceeding 100 W, fed from a readily accessible socket outlet, are exempt from these regulations.

Exercises

1 The illuminance directly below a light source of 1000 cd suspended 5 m above the surface will be equal to:
(a) 20 lx, (b) 40 lx, (c) 100 lx, (d) 200 lx.

2 A surface is illuminated to 125 Lux by a light source suspended 4 m directly above the surface. The luminous intensity of the light source will be equal to:
(a) 7.8128 cd (b) 31.25 cd, (c) 2000 cd, (d) 3906.25 cd.

3 The efficacy of a light source is measured in terms of the rate at which electricity can be converted to light. The efficacy of a light source is measured in:
(a) amperes per candela, (b) Lux per watt, (c) volts per candela, (d) lumens per watt.

4 The light output of a luminaire is reduced during its life because of the accumulation of dust and dirt on the luminaire and the interior decorations. The factor which takes this into account is called the:

(a) illuminance factor, (b) utilisation factor, (c) maintenance factor, (d) depreciation factor.

5 A light source which emits the whole range of wavelengths corresponding to visible radiation can be said to have:
(a) good colour rendering properties, (b) poor colour rendering properties, (c) good efficacy, (d) poor efficacy

6 Lamps which produce light as a result of the excitation of a gas or metallic vapour are known as:
(a) general lighting service lamps, (b) discharge lamps, (c) filament lamps, (d) incandescent lamps.

7 A fluorescent tube can be more accurately described as:
(a) an incandescent lamp, (b) a low pressure sodium lamp, (c) a high pressure mercury vapour lamp, (d) a low pressure mercury vapour lamp.

8 The rating of the final circuits for discharge lamps may be taken as the rated lamp watts multiplied by a factor of:
(a) 0.75, (b) 1.5, (c) 1.8, (d) 2.0.

9 If eight 60 W fluorescent luminaires were connected to one final circuit, the assumed rating of the circuit would be:
(a) 7.5 W, (b) 48 W, (c) 480 W, (d) 864 W.

10 The secondary voltage of the transformer used for high voltage discharge lighting must not exceed a voltage to earth of:
(a) 55 V, (b) 110 V, (c) 5 kV, (d) 10 kV.

11 Use a sketch to describe the construction and operation of a tungsten halogen lamp. State the lamps efficacy, rated life, burning position and colour rendering properties.

12 Use a sketch to describe the construction and operation of a high pressure mercury vapour lamp. State the lamps efficacy, rated life, burning position and colour rendering properties.

13 Use a sketch to describe the construction and operation of a low pressure sodium lamp. State the lamps efficacy, rated life, burning position and colour rendering properties.

14 Use a sketch to describe the construction and operation of a high pressure sodium lamp. State the efficacy, rated life, burning position and colour rendering properties of these lamps.

15 Sketch the circuit diagram and describe the operation of
(a) a switch-start fluorescent lamp circuit, (b) a quick start fluorescent lamp circuit, (c) a semi-resonant start fluorescent lamp circuit.

16 Use a block diagram to describe the control isolation and protection equipment to be considered when installing a high voltage discharge lighting circuit.

17 State one typical application for each of the following lamps:
(a) GLS lamps, (b) tungsten halogen lamps, (c) low pressure mercury vapour lamps, (d) high pressure mercury vapour lamps, (e) low pressure sodium lamps, (f) high pressure sodium lamps.

18 Explain why the light output from a high pressure sodium lamp does not reach a maximum value until some time after the lamp is switched on.

19 A high pressure mercury vapour lamp takes some minutes to restrike when it is switched off and immediately switched on again. Explain the reason for this and the limitations which this places on such a lamp.

20 A large machine shop is to be illuminated by discharge lamps. Explain what is meant by the stroboscopic effect and how it can be eliminated in a machine shop.

21 A street lantern suspends a 3000 cd high pressure sodium lamp 5 m above the ground. Determine the illuminance directly below the lamp and at a distance of 5 m from the base of the lantern.

22 A workshop measuring 12 m × 8 m is to be illuminated to an illuminance of 300 Lux. The electrical specification requires that 1500 mm fluorescent luminaires be used, each having an output of 3 900 lumens. Calculate the number of luminaires required for this installation when the maintenance factor and utilisation factor are respectively 0.9 and 0.7.

Draw to scale a plan of the workshop and show a suitable layout for the luminaires.

CHAPTER 10

Electronics

The use of electronic circuits in electrical installation work has increased considerably over recent years. Electronic circuits and components can now be found in motor starting and control circuits, discharge lighting, emergency lighting, alarm circuits and special effects lighting systems. There is therefore a need for the installation electrician to become familiar with some basic electronics which is the aim of this chapter and Chapter 11 of *Basic Electrical Installation Work*.

There are thousands of electronic components, diodes, transistors, thyristors and integrated circuits each with their own limitations, characteristics and designed application. When repairing electronic circuits it is important to replace damaged components with an identical or equivalent component. Manufacturers issue comprehensive catalogues with details of working voltage, current, power dissipation etc., and the reference numbers of equivalent components. These catalogues, together with a high impedance multimeter, should form a part of the extended tool kit for an installation electrician proposing to repair electronic circuits.

Test instruments

Electrical installation circuits usually carry in excess of one ampere and often carry hundreds of amps. Electronic circuits operate in the milliampere or even micro-ampere range. The test instruments used on electronic circuits must have a high impedance so that they do not damage the circuit when connected to take readings.

All instruments cause some disturbance when connected into a circuit because they consume some power in order to provide the torque required to move the pointer. In power applications these small disturbances seldom give rise to obvious errors, but in electronic circuits a small disturbance can completely invalidate any readings taken. Consider a voltmeter of resistance 100 kΩ connected across the circuit shown in Figure 10.2(a).

Connection of the meter loads the circuit by effectively connecting a 100 kΩ resistor in parallel with the circuit resistor as shown in Figure 10.2(b) which changes the circuit to that shown in Figure 10.2(c).

Common sense tells us that the voltage across each resistor will be 100 volts but the meter would read about 66 volts because connection of the meter has changed the circuit. This loading effect can be reduced by choosing instruments which have a very high impedance. Such an instrument imposes less load on the circuit and gives an indication much closer to the true value.

Commercial multi-range instruments reading volts, amps and ohms are usually the most convenient test instrument for an electrician, although a cathode ray oscilloscope (CRO) can be invaluable for bench work.

Commercial ohm meter

When using a commercial multi-range meter, such as an ohm meter, for testing electronic components care must be exercised in identifying the positive terminal. The red terminal of the meter, identifying the positive input for testing voltage and current, usually becomes the negative terminal when used as an ohm meter because of the way the internal battery is connected to the meter movement. To reduce confusion when using a multi-range meter as an ohm meter it is advisable

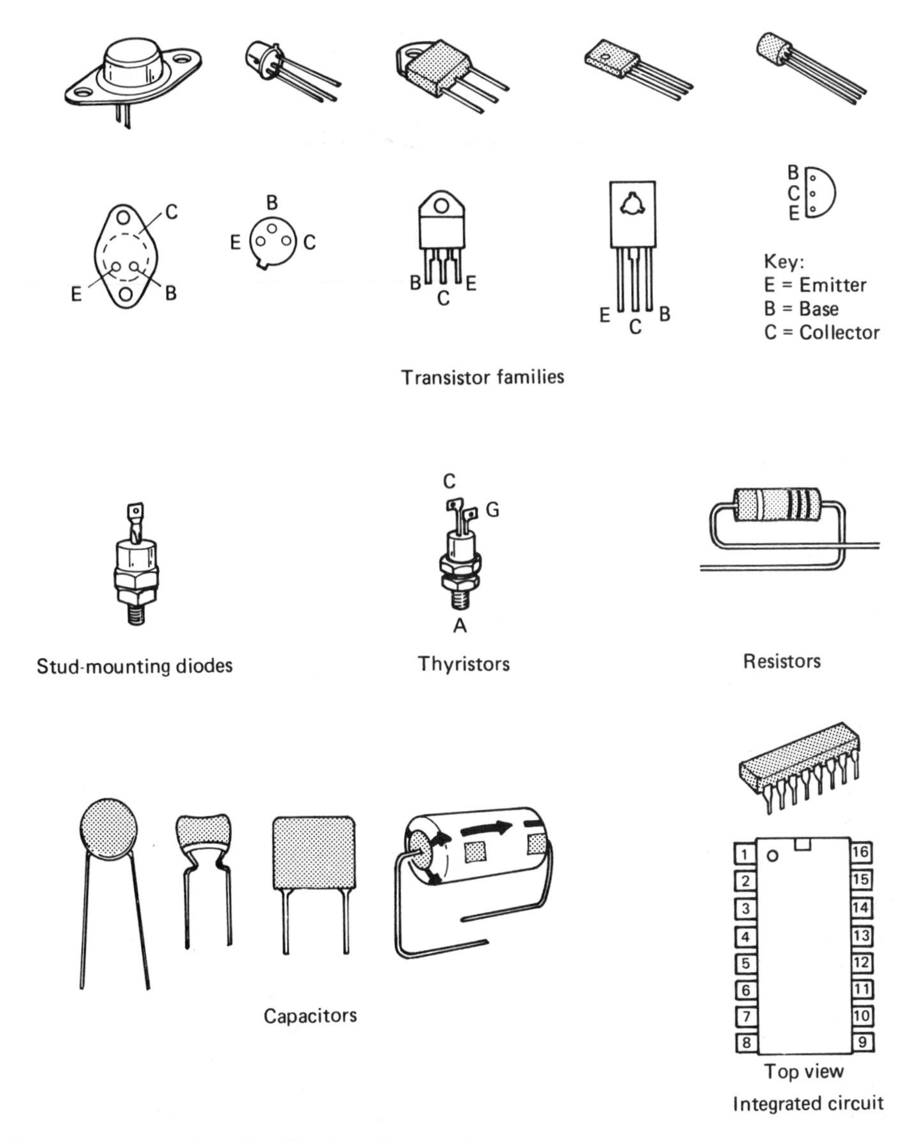

Figure 10.1 The appearance and pin identification of some electronic components

to connect the red lead to the black terminal and the black lead to the red terminal so that the red lead indicatives positive and the black lead negative as shown in Figure 10.3

The cathode ray oscilloscope

The CRO is a high impedance voltmeter which allows us to 'look into' a circuit and 'see' the waveforms present. Because of a calibrated y-axis and the 1 cm grating on the tube front we can also measure the waveform, provided that we first set the variable controls to calibrate.

Use of the CRO to measure voltage and frequency

The brilliance and focus controls should be adjusted so that a sharp image is formed which is not too bright, otherwise the fluorescent phosphor

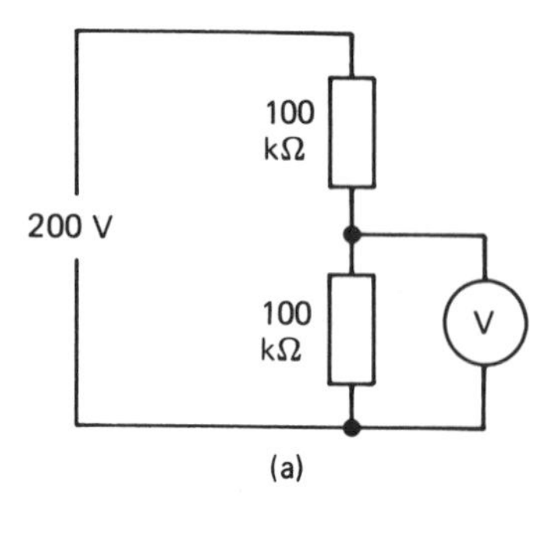

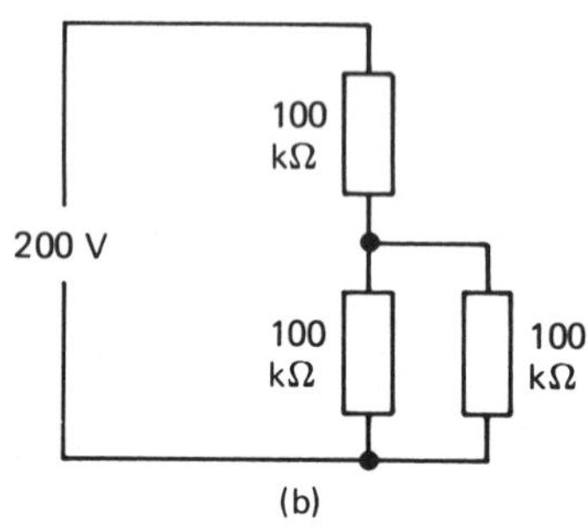

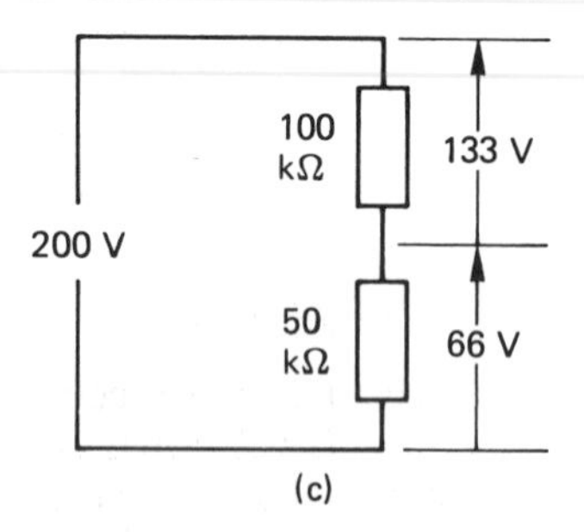

Figure 10.2 Circuit disturbance caused by the connection of a voltmeter

on the screen may become damaged. Adjust controls to calibrate and connect the test signal to the y input. Adjust the x and y tuning controls until a steady trace is obtained on the screen such as that shown in Figure 10.4.

To measure the voltage of the signal shown in Figure 10.4, count the number of centimetres from one peak of the waveform to the other using the centimetre grating. This distance is shown as 4 cm in Figure 10.4. This value is then multiplied by the volts/cm indicated on the y amplifier control knob. If the knob was set to say 2 V/cm, the peak to peak voltage of Figure 10.4 would be 4 cm × 2 V/cm = 8 V. The peak voltage would be 4 V and the rms voltage 0.7071 × 4 = 2.828 V.

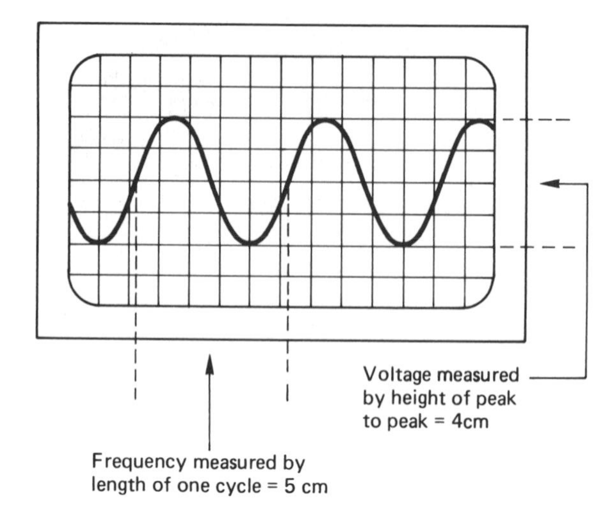

Figure 10.4 Typical trace on a CRO screen

To measure the frequency of the waveform shown in Figure 10.4, count the number of centimetres for one complete cycle using the one centimetre grating. The distance is shown as 5 cm in Figure 10.4. This value is then multiplied by the time/cm on the x amplifier or time base amplifier control knob. If this knob were set to 4 ms/cm the time taken to complete one cycle would be 5 cm × 4 ms/cm = 20 ms. Frequency can be found from

$$f = \frac{1}{T}\,(\text{Hz}),$$

and therefore $f = \dfrac{1}{20 \times 10^{-3}} = \dfrac{1000}{20} = 50$ Hz.

The waveform shown in Figure 10.4 therefore has an rms voltage of 2.828 V at a frequency of 50 Hz. The voltage and frequency of any waveform can be found in this way.

The junction diode

This is the most basic of all semiconductor devices being made from p-type material fused to n-type material as described in Chapter 11 of *Basic Electrical Installation Work*. The p-n junction of the junction diode has a low resistance in one direction and a very high resistance in the reverse direction.

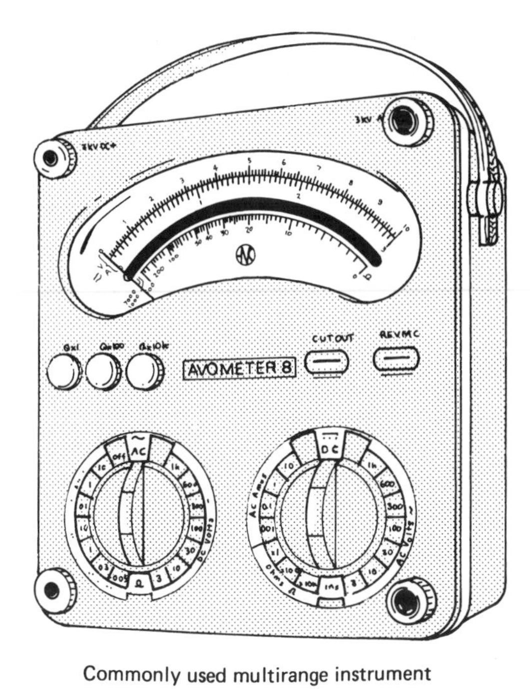

Commonly used multirange instrument

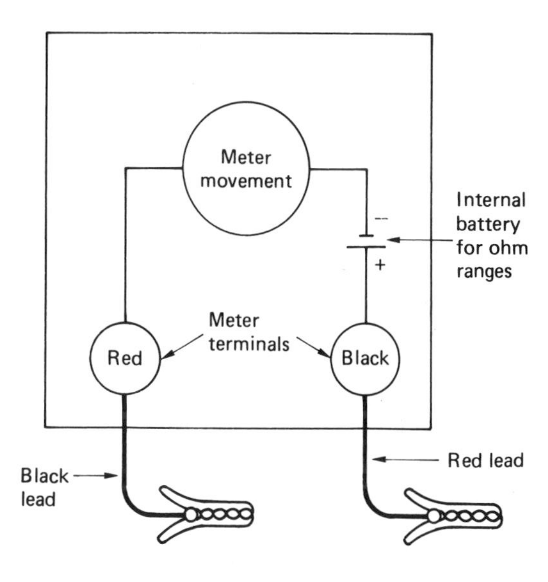

Figure 10.3 Multi-range meter used as an ohm meter (by kind permission of G. Waterworth and R. P. Phillips)

Connecting an ohm meter as described in Figure 10.3 with the red positive lead to the anode of the junction diode and the black negative lead to the cathode of the junction diode, would give a very low resistance reading. Reversing the lead connections would give a high resistance reading. This test can be used to identify a 'good' component.

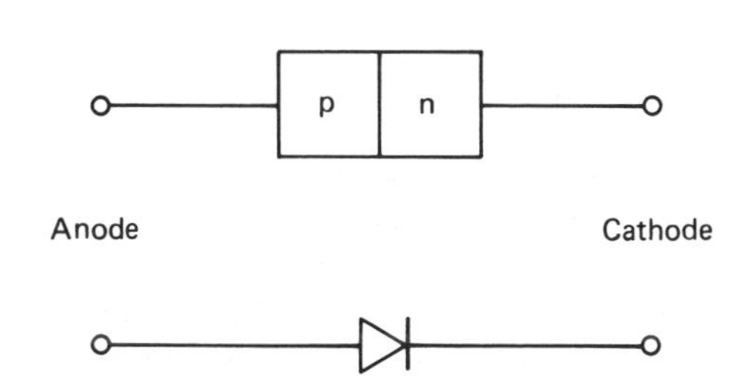

Figure 10.5 A junction diode

The bi-polar transistor

The transistor is the modern semiconductor equivalent of the thermionic valve and is the simplest component which can be used to amplify a signal. The transistor consists of three pieces of semiconductor material sandwiched together as shown in Figure 10.6.

The structure of the transistor makes it a three terminal device having a base, collector and emitter terminal. By varying the current flowing into the base connection a much larger current flowing between collector and emitter can be controlled. Apart from the supply connections, both n-p-n and p-n-p types are essentially the same but the n-p-n type is more common.

Transistor testing

A transistor can be thought of as two diodes connected together. Therefore, a transistor can be tested using an ohm meter in the same way that was described for the diode.

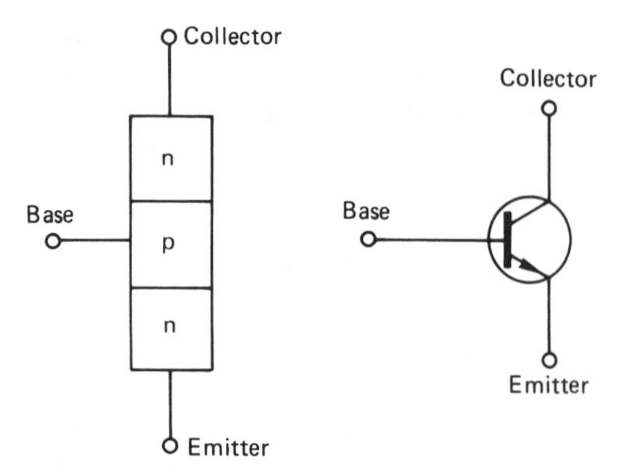

(a) Structure and symbol of n-p-n transistor

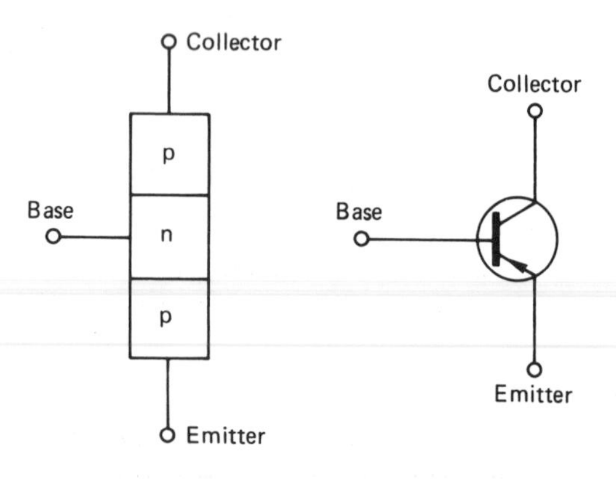

(b) Structure and symbol of p-n-p transistor

Figure 10.6 n-p-n and p-n-p bi-polar transistor

Table 10.1 Transistor testing using an Ohm Meter

A 'good' n–p–n transistor will give the following readings	
Red to base and black to collector	= low resistance
Red to base and black to emitter	= low resistance
Reversed connections on the above terminals will result in a high resistance reading as will connections of either polarity between the collector and emitter terminals.	

A 'good' p–n–p transistor will give the following readings	
Black to base and red to collector	= low resistance
Black to base and red to emitter	= low resistance
Reversed connections on the above terminals will result in a high resistance reading as will connections of either polarity between the collector and emitter terminals.	

Assuming that the red lead of the ohm meter is positive, as described in Figure 10.3, the transistor can be tested in accordance with Table 10.1. When many transistors are to be tested a simple test circuit can be assembled as shown in Figure 10.7.

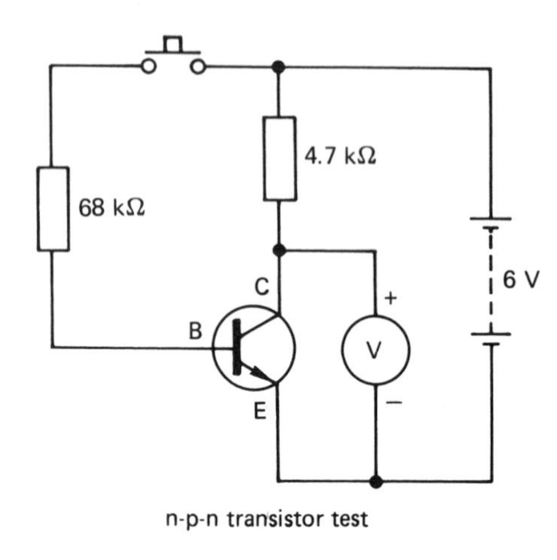

n-p-n transistor test

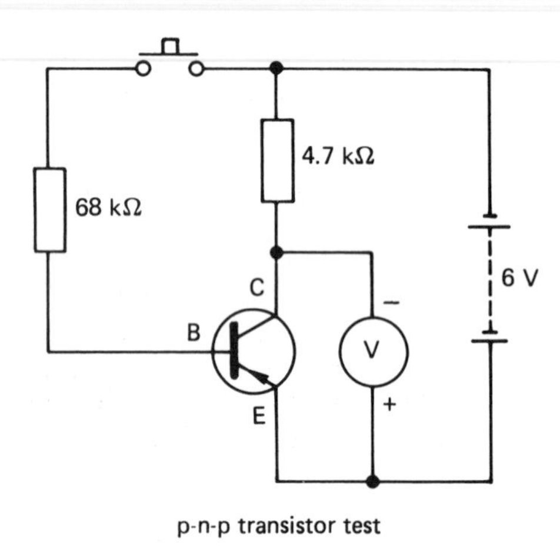

p-n-p transistor test

Figure 10.7 Transistor testing using an ohm meter

With the circuit connected as shown in Figure 10.7, a 'good' transistor will give readings on the voltmeter of 6 V with the switch open and about 0.5 V when the switch is closed. The voltmeter used for the test should have a high internal resistance, about ten times greater than 4.7 kΩ; this is usually indicated on the back of a multi-range meter or in the manufacturer's information supplied with a new meter.

The thyristor or silicon-controlled rectifier (SCR)

The thyristor consists of four pieces of semiconductor material sandwiched together and connected to three terminals as shown in Figure 10.8.

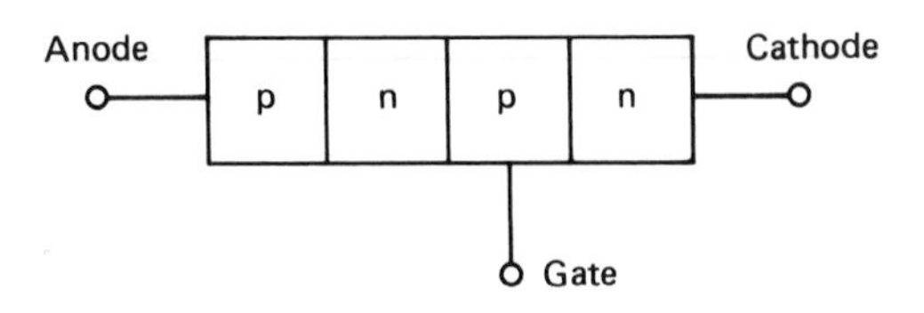

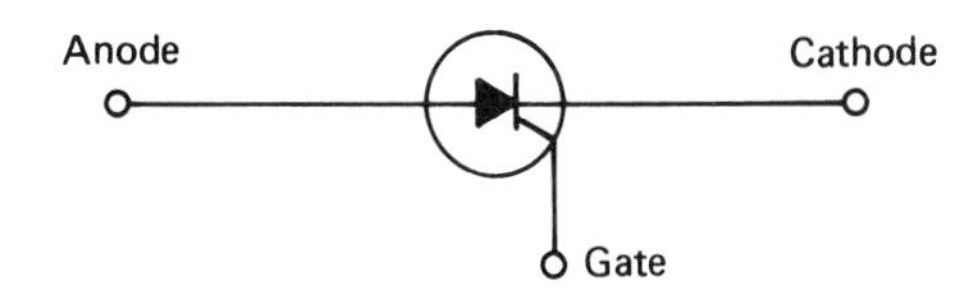

Figure 10.8 Structure and symbol of a thyristor

The thyristor behaves like a door: It can be open or shut, allowing or preventing current to flow through the device. The door is opened, or we say the thyristor is triggered, to a conducting state by applying a pulse voltage to the gate connection. Once the thyristor is in the conducting state, the gate loses all control over the device. The only way to bring the thyristor back to a non-conducting state is to reduce the voltage across the anode and cathode to zero or apply reverse voltage across the anode and cathode. With an a.c. supply connected this occurs every half cycle. Figure 10.9. shows

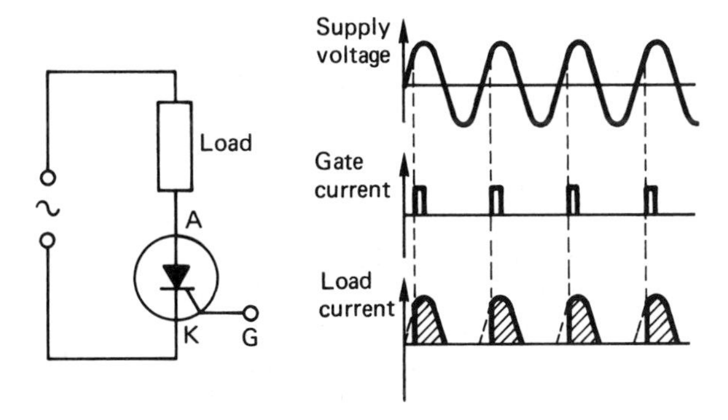

Figure 10.9 Waveforms to show the control effect of a thyristor

that by varying the point at which the gate signal triggers the thyristor, the output waveform and the power available at the load can be varied. Power is reduced by triggering the gate later in the cycle.

Thyristor in practice

The thyristor has no moving parts and operates without arcing. It can operate at extremely high speeds and the currents used to operate the gate are very small. The most common application for the thyristor is to control the power supply to a load, for example, lighting dimmers. Some simple control circuits and the resultant waveforms are shown at Figure 10.10. The trigger circuits required to operate the gate G have been excluded for simplicity and clarity.

Thyristor testing using an ohm meter

A thyristor can be tested with an ohm meter as described in Table 10.2, assuming that the red lead of the ohm meter is positive as described in Figure 10.3.

Table 10.2 Thyristor Testing using an Ohm Meter

A 'good' thyristor will give the following readings
Black to cathode and red on gate = low resistance Red to cathode and black on gate = a higher resistance value
The value of the second reading will depend upon the thyristor, and may vary from only slightly greater to very much greater.
From cathode to anode with either polarity connected will result in a very high resistance reading.

Thyristor testing

A simple test circuit is shown in Figure 10.11. When SW.B only is closed the lamp will not light, but when SW.A is also closed, the lamp lights to full brilliance. The lamp will remain illuminated even when SW.A is opened. This shows that the thyristor is operating correctly: Once a voltage has been applied to the gate the thyristor becomes forward conducting, like a diode, and the gate loses control.

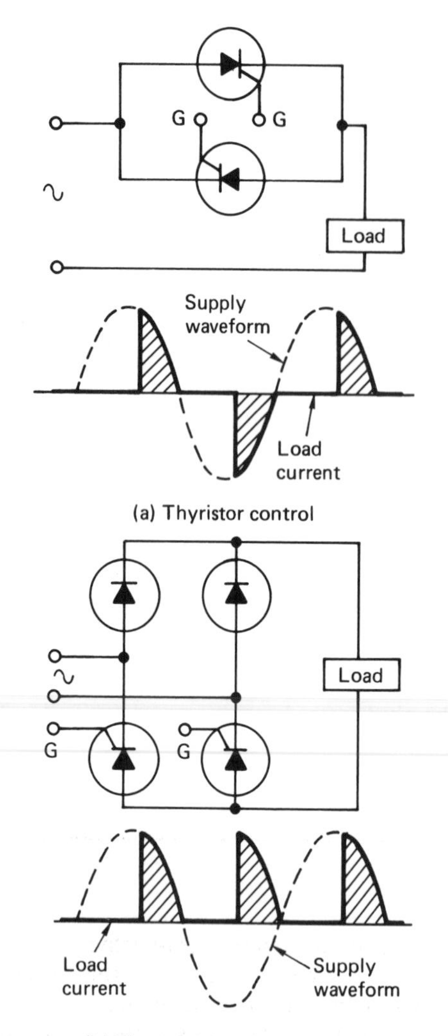

Figure 10.10 Simple thyristor control circuits

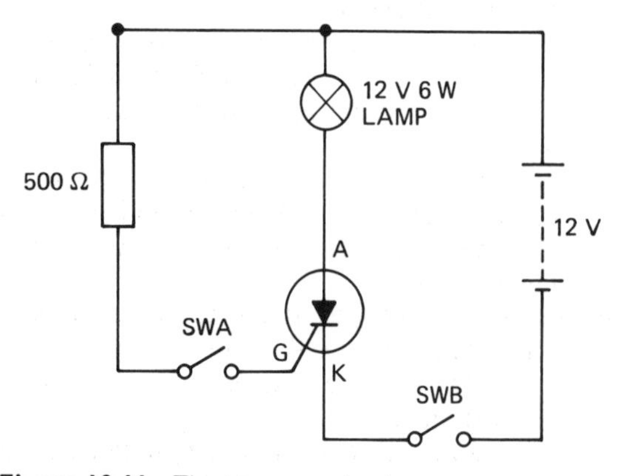

Figure 10.11 Thyristor test circuit

Integrated circuits

Modern electronics relies on transistors. They are made from silicon and can be manufactured efficiently, cheaply and with good reliability. They are the basic building blocks of all electronic circuits. When several or even hundreds of transistors are etched onto a single wafer of silicon and connected together in a single package we call the device an integrated circuit. The integrated circuit or IC is the electronic revolution. ICs are more reliable, cheaper, smaller and electronically superior to the same circuit made from discreet or separate transistors. One IC behaves differently to another because of the arrangement of the transistors within the IC (see Figure 10.12).

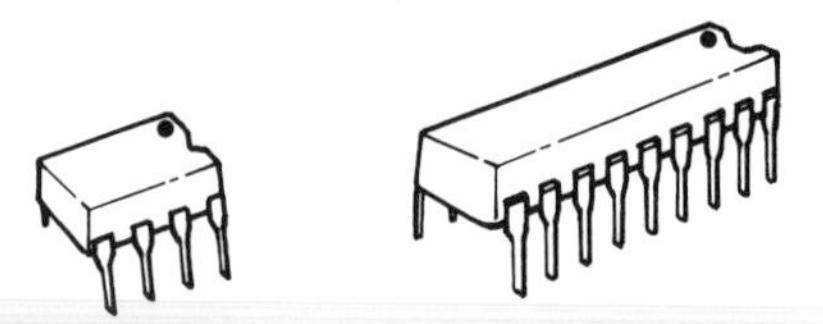

Figure 10.12 DIL packaged integrated circuits

Manufacturers' data sheets describe the characteristics of the different ICs which have a reference number stamped on the top. When building circuits it is necessary to be able to identify the IC pin connection by number. The number one pin of any IC is indicated by a dot pressed into the encapsulation or is the pin to the left of the cut out as shown in Figure 10.13. Since the packaging of ICs has two rows of pins they are called DIL (dual in line) packaged integrated circuits.

Integrated circuits are sometimes connected into DIL sockets and at other times are soldered directly into the circuit. The testing of ICs is beyond the scope of a practising electrician and when they are suspected of being faulty an identical or equivalent replacement should be connected into the circuit, ensuring that it is inserted the correct way round, which is indicated by the position of pin number one as described earlier.

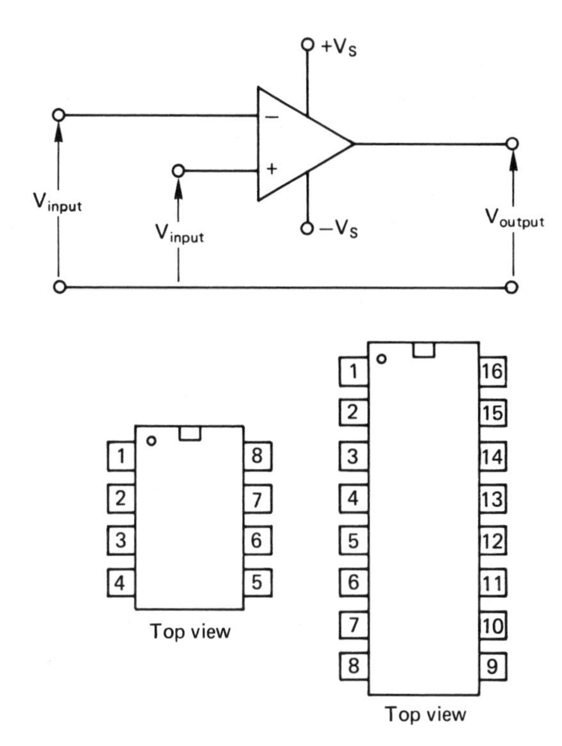

Figure 10.13 IC amplifier symbol and pin identification

Exercises

1 Since electronic circuits operate in the milli-ampere or micro-ampere range, it is important that the test instruments used on electronic circuits have:
(a) a high impedance, (b) a low impedance, (c) an analogue scale, (d) a digital readout.

2 A sinusoidal waveform is displayed on the screen of a CRO. The controls on the y axis are set to 10 V/cm and the measurement from peak to peak is measured as 4 cm. The rms value of the waveform will be:
(a) 7.07 V, (b) 14.14 V, (c) 28.28 V, (d) 40 V.

3 A sinusoidal waveform is displayed on the screen of a CRO. The controls on the x axis are set to 2 ms/cm and the measurement for one period is calculated to be 5 cm. The frequency of the waveform will be:
(a) 10 Hz, (b) 50 Hz, (c) 100 Hz, (d) 10 kHz.

4 A junction diode is connected into a simple circuit. When the anode of the diode is positive with respect to the cathode:
(a) a current will flow in the circuit, (b) no current can flow in the circuit, (c) the output will be amplified, (d) reverse breakdown voltage will occur.

5 Voltages as shown below are connected across the anode and cathode of a junction diode. The diode will conduct when:
(a) the anode = 2 V and cathode = 4 V, (b) the anode = 0 V and cathode = 2 V, (c) the anode = − 4 V and cathode = − 2 V, (d) the anode = − 2 V and cathode = − 4 V.

6 A thyristor can be triggered by applying a pulse voltage to the thyristor:
(a) base, (b) gate, (c) anode, (d) cathode.

7 When a thyristor is conducting it can only be brought back to a non-conducting state:

(a) by applying a positive potential to the cathode, (b) by controlling the gate voltage, (c) by controlling the gate current, (d) by reducing the voltage across the anode and cathode to zero.

8 The pin connections of a dual in line integrated circuit are numbered by an agreed convention. When viewed from the top of the IC pin number one is always the pin:
(a) to the right of the U–shaped cut out, (b) to the left of the U-shaped cut out, (c) diagonally opposite the U-shaped cut out, (d) parallel with the U-shaped cut out.

9 Describe with the aid of a sketch a method of determining the rms value of an a.c. sinusoidal waveform displayed on the screen of a CRO and measuring 5 cm on the screen grating from one peak to the other.

10 With the aid of a sketch describe a step by step method of determining the frequency of a sinusoidal waveform displayed on the screen of a CRO and measuring 8 cm for one period of the waveform.

11 Use a sketch to describe how a junction diode may be tested for correct operation using
(a) an ohm meter and (b) a simple circuit.

12 Describe how an n-p-n transistor may be tested using
(a) an ohm meter and (b) a simple test circuit.
Indicate the results to be anticipated for a 'good' transistor.

13 Describe how a thyristor may be used to control the power available at a load.

14 Explain one method of testing a thyristor using
(a) an ohm meter and (b) a simple test circuit.
Clearly indicate the test results for a 'good' thyristor.

15 What is meant by 'an integrated circuit'. Describe how the pin numbers are identified for a DIL packaged IC.

CHAPTER 11

Management skills

Smaller electrical contracting firms will know where their employees are working and what they are doing from day to day because of the level of personal contact between the employer, employee and customer.

When firms expand and become engaged on larger contracts, the level of total knowledge held by any one individual diminishes, and there becomes an urgent need for sensible management and planning skills so that men and materials are on site when they are required and a healthy profit margin is maintained.

When the electrical contractor is told that he has been successful in tendering for a particular contract he is committed to carry out the necessary work within the contract period. He must therefore consider:

- by what date the job must be finished,
- when the job must be started if the completion date is not to be delayed,
- how many men will be required to complete the contract,
- when certain materials will need to be ordered,
- when the supply authorities must be notified that a supply will be required,
- if it is necessary to obtain authorisation from a statutory body for any work to commence.

In thinking ahead and planning the best method of completing the contract, the individual activities or jobs must be identified and consideration given to how the various jobs are interrelated. To help in this process a number of management techniques are available. In this chapter we will consider only two; bar charts and network analysis. The very preparation of a bar chart or network analysis forces the contractor to think deeply, carefully and logically about the particular contract, and it is therefore a very useful aid to the successful completion of the work.

Bar charts

There are many different types of bar chart used by industry, but the object of any bar chart is to establish the sequence and timing of the various activities involved in the contract as a whole. They are a visual aid in the process of communication. In order to be useful they must be clearly understood by the people involved in the management of a contract. The chart is constructed on a rectangular basis as shown in Figure 11.1.

All the individual jobs or activities which make up the contract are identified and listed separately down the vertical axis on the left hand side, and time flows from left to right along the horizontal axis. The unit of time can be chosen to suit the length of the particular contract, but for most practical purposes either days or weeks are used.

The simple bar chart shown in Figure 11.1(a) shows a particular activity A which is estimated to last two days, while activity B lasts eight days. Activity C lasts four days and should be started on day three. The remaining activities can be interpreted in the same way.

With the aid of colours, codes, symbols and a little imagination, much additional information can be included on this basic chart. For example, the actual work completed can be indicated by shading above the activity line as shown in Figure 11.1(b) with a vertical line indicating the number of contract days completed, the activities which are on time, ahead or behind time can easily be identified. Activity B in Figure 11.1(b) is two days behind schedule, while activity D is two days

Time / Activity	Day number													
	1	2	3	4	5	6	7	8	9	10	11	12	13	14
A														
B														
C														
D														
E														
F														
G														
H														
I														
J														
K														
L														
M														

(a) A simple bar chart

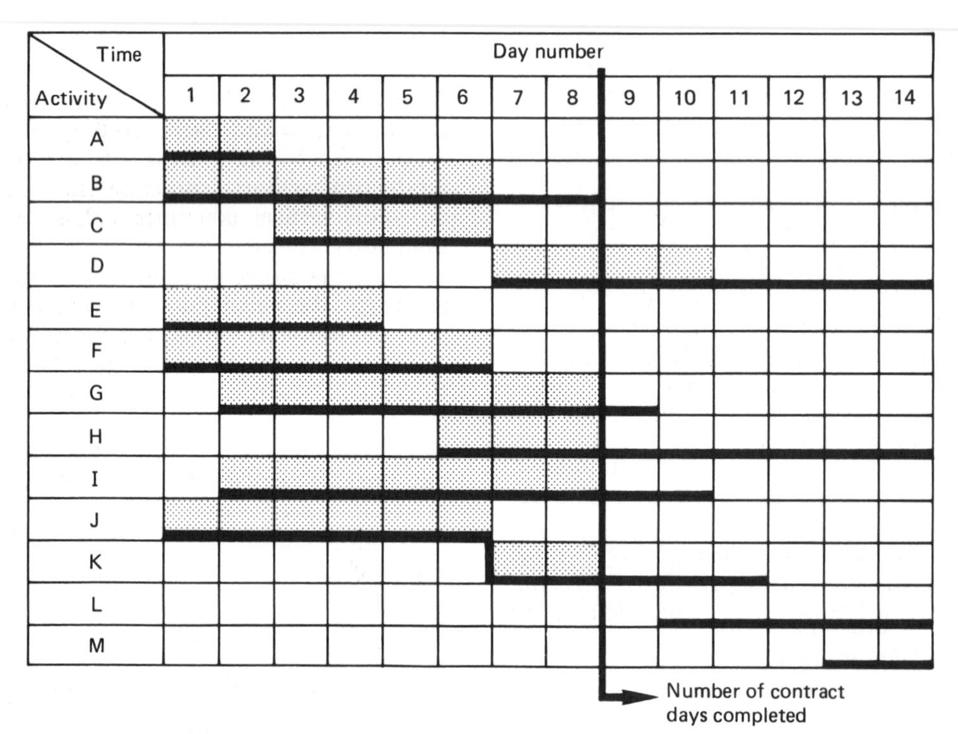

(b) A modified bar chart indicating actual work completed

Figure 11.1 Bar charts

ahead of schedule. All other activities are on time. Some activities must be completed before others can start. For example, all conduit work must be completely erected before the cables are drawn in. This can be shown in Figure 11.1(b) by activities J and K. The short vertical line between the two activities indicates that activity J must be completed before K can commence.

Useful and informative as the bar chart is, there is one aspect of the contract which it cannot display. It cannot indicate clearly the interdependence of the various activities upon each other, and it is unable to identify those activities which must strictly adhere to the time schedule if the overall contract is to be completed on time, and those activities in which some flexibility is acceptable. To overcome this limitation, in 1959 the CEGB developed the critical path network diagram which we will now consider.

Network analysis

In large or complex contracts there are a large number of separate jobs or activities to be performed. Some can be completed at the same time, whilst others cannot be started until others are completed. A network diagram can be used to coordinate all the interrelated activities of the most complex project in such a way that all sequential relationships between the various activities, and the restraints imposed by one job on another, are allowed for. It also provides a method of calculating the time required to complete an individual activity and will identify those activities which are the key to meeting the completion date, called the critical path. Before considering the method of constructing a network diagram, let us define some of the terms and conventions we shall be using.

Critical path

This is the path taken from the start event to the end event which takes the longest time. This path denotes the time required for completion of the whole contract.

Float time

Slack time or time in hand is the time remaining to complete the contract after completion of a particular activity.

Float time = Critical path time − Activity time.

The total float time for any activity is the total leeway available for all activities in the particular path of activities in which it appears. If the float time is used up by one of the early activities in the path, there will be no float left for the remaining activities and they will become critical.

Activities

Activities are represented by an arrow, the tail of which indicates the commencement, and the head the completion of the activity. The length and direction of the arrows have no significance; they are not vectors or phasors. Activities require time, manpower and facilities. They lead up to or emerge from events.

Dummy activities

Dummy activities are represented by an arrow with a dashed line. They signify a logical link only, require no time and denote no specific action or work.

An event

An event is a point in time, a milestone or stage in the contract when the preceding activities are finished. Each activity begins and ends in an event. An event has no time duration and is represented by a circle which sometimes includes an identifying number or letter.

Time may be recorded to a horizontal scale or shown on the activity arrows. For example, the activity from event A to B takes 9 hours in the network diagram shown in Figure 11.2.

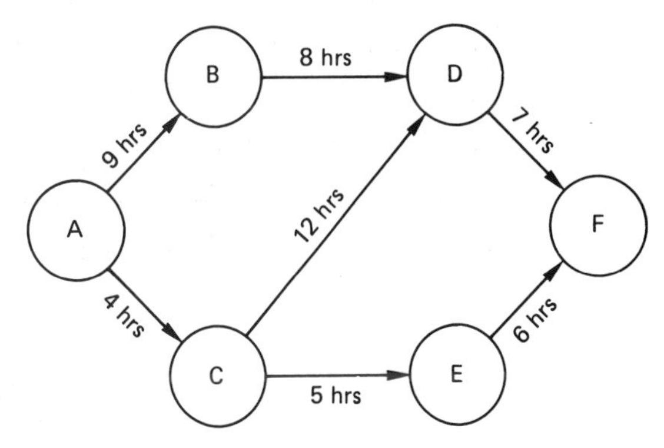

Figure 11.2 A network diagram for Example 1

Example 1

Identify the three possible paths from the start event A to the finish event F for the contract shown by the network diagram in Figure 11.2. Identify the critical path and the float time in each path.

The three possible paths are:

1 event A–B–D–F
2 event A–C–D–F
3 event A–C–E–F

The times taken to complete these activities are:

1 path A–B–D–F = 9 + 8 + 7 = 24 hours
2 path A–C–D–F = 4 + 12 + 7 = 23 hours
3 path A–C–E–F = 4 + 5 + 6 = 15 hours

The longest time from the start event to the finish event is 24 hours, and therefore the critical path is A–B–D–F.

Float time = Critical path – Activity time.

For path 1, A–B–D–F,
Float time = 24 hours – 24 hours = 0 hours.

There can be no float time in any of the activities which form a part of the critical path since a delay on any of these activities would delay completion of the contract. On the other two paths some delay could occur without affecting the overall contract time.

For path 2, A–C–D–F,
Float time = 24 hours – 23 hours = 1 hour.

For path 3, A–C–E–F,
Float time = 24 hours – 15 hours = 9 hours.

Example 2

Identify the time taken to complete each activity in the network diagram shown in Figure 11.3. Identify the three possible paths from the start event A to the final event G and state which path is the critical path.

Time taken to complete each activity using the horizontal scale

activity A–B = 2 days
activity A–C = 3 days
activity A–D = 5 days
activity B–E = 5 days
activity C–F = 5 days
activity E–G = 3 days
activity D–G = 0 days
activity F–G = 0 days

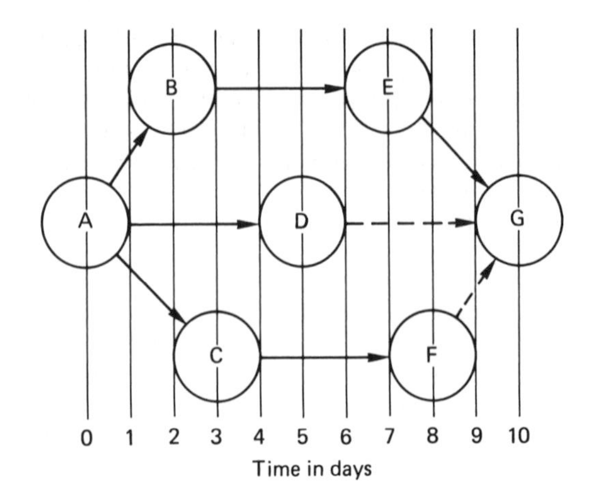

Figure 11.3 A network diagram for Example 2

Activities D–G and F–G are dummy activities which take no time to complete but indicate a logical link only. This means that in this case once the activities preceding events D and F have been completed, the contract will not be held up by work associated with these particular paths and they will progress naturally to the finish event.

The three possible paths are:

1 A–B–E–G
2 A–D–G
3 A–C–F–G

The times taken to complete the activities in each of the three paths are:

Path 1, A–B–E–G = 2 + 5 + 3 = 10 days
Path 2, A–D–G = 5 + 0 = 5 days
Path 3, A–C–F–G = 3 + 5 + 0 = 8 days

The critical path is path 1, A–B–E–G

Constructing a network

The first step in constructing a network diagram is to identify and draw up a list of all the individual jobs, or activities as they are called, which require time for their completion and which must be completed to advance the contract from start to completion.

The next step is to build up the arrow network showing schematically the precise relationship of the various activities between the start and end event. The designer of the network must ask these questions:

1 Which activities must be completed before others can commence? These activities are then drawn in a similar way to a series circuit but with event circles instead of resistor symbols.
2 Which activities can proceed at the same time? These can be drawn in a similar way to parallel circuits but with event circles instead of resistor symbols.

Commencing with the start event at the left hand side of a sheet of paper, the arrows representing the various activities are built up step by step until the final event is reached. A number of attempts may be necessary to achieve a well balanced and symmetrical network diagram showing the best possible flow of work and information, but this time is well spent when it produces a diagram which can be easily understood by those involved in the management of the particular contract.

Example 3

A particular electrical contract is made up of activities A to F as described below:

A = an activity taking 2 weeks commencing in week 1
B = an activity taking 3 weeks commencing in week 1
C = an activity taking 3 weeks commencing in week 4
D = an activity taking 4 weeks commencing in week 7
E = an activity taking 6 weeks commencing in week 3
F = an activity taking 4 weeks commencing in week 1

Certain restraints are placed on some activities because of the availability of men and materials and the constraint that some work must be completed before other work can commence as follows:

Activity C can only commence when B is completed.
Activity D can only commence when C is completed.
Activity E can only commence when A is completed.
Activity F does not restrict any other activity.

(a) Produce a simple bar chart to display the activities of this particular contract.
(b) Produce a network diagram of the programme and describe each event.
(c) Identify the critical path and the total contract time.
(d) State the maximum delay which would be possible on activity E without delaying the completion of the contract.
(e) State the float time in activity F.

Time / Activities	Week numbers									
	1	2	3	4	5	6	7	8	9	10
A										
B										
C										
D										
E										
F										

(a) Bar chart for Example 3

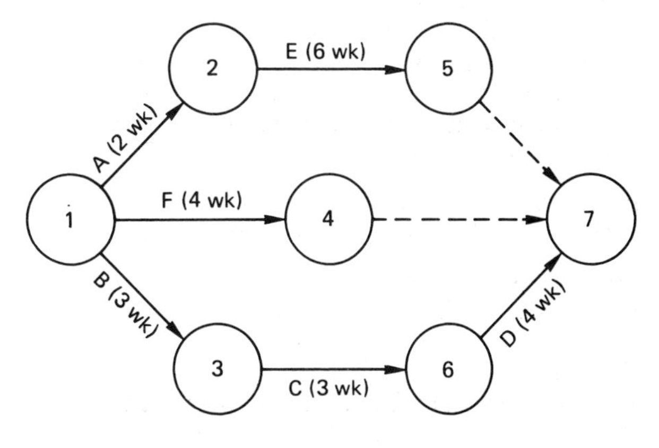

(b) Network diagram for Example 3

Figure 11.4 Bar chart and network diagram

(a) A simple bar chart for this contract is shown in Figure 11.4(a)
(b) The network diagram is shown in Figure 11.4(b).

Event 1 = the commencement of the contract.
Event 2 = the completion of activity A and the commencement of activity E.
Event 3 = the completion of activity B and the commencement of activity C

Event 4 = the completion of activity F
Event 5 = the completion of activity E
Event 6 = the completion of activity C
Event 7 = the completion of activity D and the whole contract.

(c) There are three possible paths:

1 via events 1–2–5–7,
2 via events 1–4–7,
3 via events 1–3–6–7.

The time taken for each path is:

path 1 = 2 weeks + 6 weeks = 8 weeks,
path 2 = 4 weeks = 4 weeks,
path 3 = 3 weeks + 3 weeks + 4 weeks = 10 weeks.

The critical path is therefore path 3, via events 1–3–6–7, and the total contract time is 10 weeks.

(d) Float time = Critical path – Activity time. time

Activity E is on path 1 via events 1–2–5–7 having a total activity time of 8 weeks.

Float time = 10 weeks – 8 weeks = 2 weeks.

Activity E could be delayed for a maximum of 2 weeks without delaying the completion date of the whole contract.

(e) Activity F is on path 2 via events 1–4–7 having a total activity time of 4 weeks.
Float time = 10 weeks – 4 weeks = 6 weeks.

Symbols

When reading or producing electrical drawings individual items of equipment are identified by graphical symbols. The standard symbols used by the electrical contracting industry are those recommended by the British Standard 3939 'Graphical symbols for electrical power, telecommunications and electronic diagrams'. Some of the more common electrical installation symbols are given in Figure 11.5.

A layout drawing is shown in Figure 11.6 of a small domestic extension. It can be seen that the mains intake position, probably a consumer's unit, is situated in the store room which also contains one light controlled by a switch at the door. The bathroom contains one lighting point controlled by a one-way switch at the door. The kitchen has two doors and a switch is installed at each door to control the fluorescent luminaire. There are also three double sockets situated around the kitchen. The sitting room has a two way switch at each door controlling the centre lighting point. Two wall lights with built in switches are to be wired, one at each side of the window. Two double sockets and one switched socket are also to be installed in the sitting room. The bedroom has two lighting points controlled independently by two one way switches at the door.

The wiring diagrams and installation procedures for all these circuits can be found in Chapters 7 and 8 of *Basic Electrical Installation Work*.

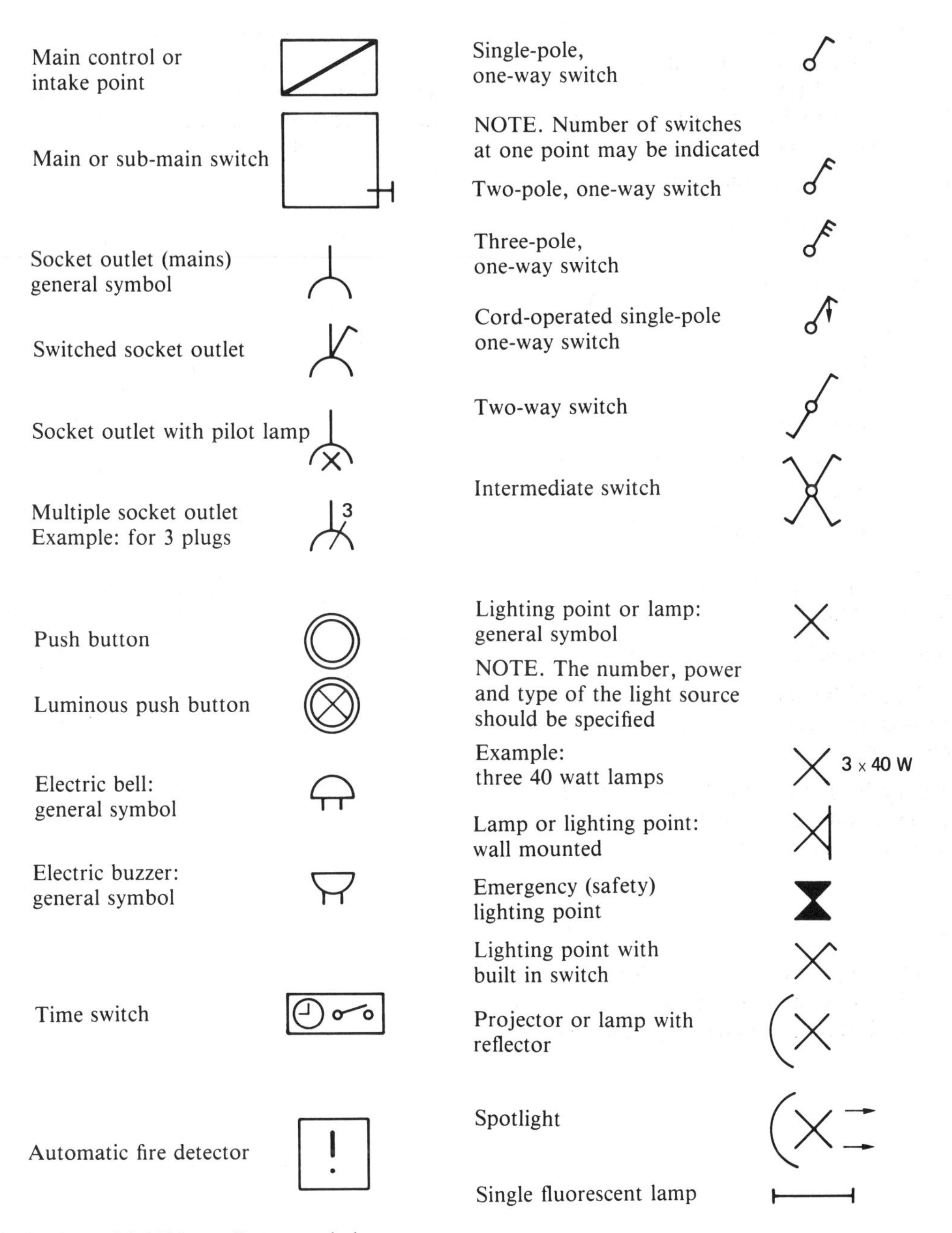

Figure 11.5 Some BS 3939 installation symbols

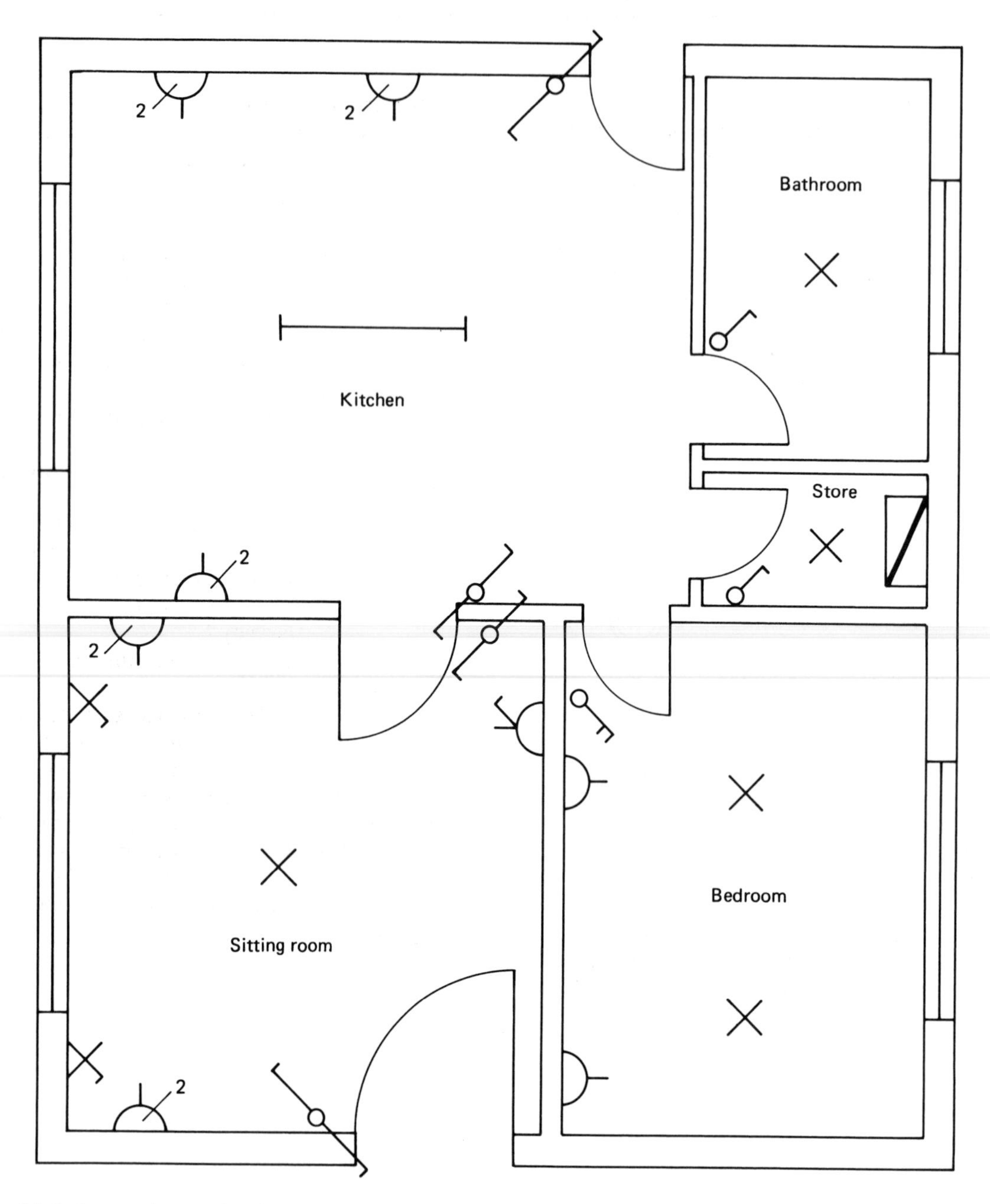

Figure 11.6 Layout drawing for electrical installation

Exercises

1 A simple bar chart can show:
(a) the activities involved in a particular contract where some flexibility is acceptable, (b) the sequence and timing of the various activities involved in a particular contract, (c) the interdependence of the various activities involved in a particular contract, (d) the total man hours involved in a particular contract.

2 A simple network diagram can show:
(a) the actual cost of a contract, (b) the actual number of man hours involved in a contract, (c) the interdependence of the various activities involved in a particular contract (d) the rating of the incoming supply cable.

3 The standard symbols used by the electrical contracting industry are those recommended by:
(a) the Institute of Electrical Engineers, (b) the Health and Safety at Work Act 1974, (c) the Factories Act 1961, (d) the British Standard 3939.

4 A particular contract is made up of activities A to I as described below:

Activity A takes 3 weeks commencing in week 1. Activity B takes 1 week commencing in week 1. Activity C takes 5 weeks commencing in week 2. Activity D takes 4 weeks commencing in week 7. Activity E takes 3 weeks commencing in week 3. Activity F takes 5 weeks commencing in week 4. Activity G takes 4 weeks commencing in week 9. Activity H takes 4 weeks commencing in week 6. Activity I takes 10 weeks commencing in week 3.

Due to the availability of men and materials some activities must be completed before others can commence as follows:

Activity C can only commence when B is completed. Activity D can only commence when C is completed. Activity F can only commence when A is completed. Activity G can only commence when F is completed. Activity H can only commence when E is completed. Activity I does not restrict any other activity.

Produce (a) a bar chart to show the various activities and
(b) a network diagram for the contract.
(c) Identify the critical path
(d) Find the time required to complete the contract
(e) State the float time in activity F.
(f) State the float time in activity D.

5 Sketch the BS 3939 graphical symbols for the following equipment: (a) A single socket outlet, (b) a double socket outlet, (c) a switched double socket outlet, (d) an electric bell. (e) a single pole one way switch, (f) a cord operated single pole one way switch, (g) a wall mounted lighting point, (h) a double fluorescent lamp, (i) an emergency lighting point.

Solutions to exercises

Chapter 1

1 b **2** d **3** Answers in text. **4** Answers in text. **5** Answers in text.

Chapter 2

1 b **2** c **3** a **4** b **5** c **6–14** Answers in text.

15	(b)	Current	A–B	= 90 A
			B–C	= 70 A
			C–D	= 40 A
	(c)	Voltage at	D	= 237V
16	(a)	Current	A–B	= 90 A
			B–C	= 50 A
			C–D	= 20 A
	(b)	Voltage at	B	= 237.075 V
			C	= 236.1 V
			D	= 235.84 V

Chapter 3

1 b **2** d **3** c **4** d **5** d **6** c **7** c **8** a **9** a **10** c **11** c **12** b **13** b **14** c **15** 1.3225 s **16** Answers in text. **17** (a) 1.14 Ω (b) 1.2 Ω (c) 4.0 Ω (d) 5.58 Ω (e) 14.1 Ω (f) 18.5 Ω **18** 5.38 mm^2 **19** (a) 0.25 s (b) 0.038 s **20** 1.466 mm^2

Chapter 4

1 c **2** b **3** a **4** c **5** d **6** c **7** Answers in text. **8** Answers in text. **9** Answers in text and Figure 4.2. **10** Answers in text. **11** Answers in text and Figure 4.3.

Chapter 5

1 a **2** b **3** a **4** a **5** d **6** c **7** c **8** a **9** b **10** d **11** (a) 18.03 Ω (b) 13.31 A (c) $V_R = 133.1$, $V_L = 199.66$ (d) 0.55 **12** (a) 15 Ω (b) 16 A (c) $V_R = 144$ V, $V_C = 192$ V (d) 0.6 **13** (a) 26.32 (b) 9.12 A (c) $V_R = 136.8$ V, $V_L = 286.55$, $V_C = 483.82$ (d) 0.569 **14** Answers in text. **15** Answers in text. **16** $I_R = 16$ A, $I_L = 7.64$ A, $I_T = 17.73$ A, p.f. = 0.902 **17** $I_R = 12$ A, $I_C = 4.52$ A, $I_T = 12.82$ A, p.f. = 0.936

Chapter 6

1 d **2** c **3** b **4** b **5** b **6** a **7–11** Answers in text. **12** 17.73 A

Chapter 7

1 d **2** b **3** c **4** b **5** a **6** c **7** b **8** d **9** b **10–22** Answers in text. **23** (a) 20.87 A (b) 0.77 μF.

Chapter 8

1 c **2** b **3** d **4** c **5** a **6** d **7** b **8** c **9** b **10** a **11–15** Answers in text. **16** (a) 7.5 m Ω, (b) 16.662 k Ω. **17–25** Answers in text.

Chapter 9

1 b **2** c **3** d **4** c **5** a **6** b **7** d **8** c **9** d **10** c **11–20** Answers in text. **21** 120 lx, 42.43 lx **22** 11.72 or 12 luminaires.

Chapter 10

1 a **2** b **3** c **4** a **5** d **6** b **7** d **8** b **9–15** Answers in text.

Chapter 11

1 b **2** c **3** d **4** (a) Answers in text, (b) Answers in text, (c) activity A–F–G, (d) 12 weeks, (e) none, (f) 2 weeks.
5 Answers in text.

INDEX